On Rheostasis

On Rheostasis

The Hierarchical Organization of Physiological Stability

TYLER JOHN STEVENSON

OXFORD
UNIVERSITY PRESS

OXFORD
UNIVERSITY PRESS

Oxford University Press is a department of the University of Oxford. It furthers the University's objective of excellence in research, scholarship, and education by publishing worldwide. Oxford is a registered trade mark of Oxford University Press in the UK and certain other countries.

Published in the United States of America by Oxford University Press
198 Madison Avenue, New York, NY 10016, United States of America.

Library of Congress Cataloging-in-Publication Data
Names: Stevenson, Tyler John, author.
Title: On rheostasis : the hierarchical organization of physiological stability / Tyler John Stevenson.
Description: New York, NY : Oxford University Press, [2024] |
Includes bibliographical references and index.
Identifiers: LCCN 2023017599 (print) | LCCN 2023017600 (ebook) |
ISBN 9780197665572 (hardback) | ISBN 9780197665596 (epub) |
ISBN 9780197665589 (ebook) | ISBN 9780197665602 (ebook other)
Subjects: LCSH: Homeostasis.
Classification: LCC QP90.4.S64 2023 (print) | LCC QP90.4 (ebook) |
DDC 612/.022—dc23/eng/20230706
LC record available at https://lccn.loc.gov/2023017599
LC ebook record available at https://lccn.loc.gov/2023017600

DOI: 10.1093/oso/9780197665572.001.0001

Printed by Integrated Books International, United States of America

MIX
Paper
FSC FSC® C183721

Contents

Introduction

واجب الرجل الذي يحقق في كتابات العلماء ، إذا كان معرفة الحقيقة هو هدفه ، هو أن يجعل نفسه عدوًا لكل ما يقرأه ، و ... يهاجمه من كل جانب. يجب عليه أيضًا أن يشك في نفسه أثناء قيامه بفحصه النقدي ، حتى يتمكن من تجنب الوقوع في التحيز أو التساهل.

> The duty of the man who investigates the writings of scientists, if learning the truth is his goal, is to make himself an enemy of all that he reads, and . . . attack it from every side. He should also suspect himself as he performs his critical examination of it, so that he may avoid falling into either prejudice or leniency.
>
> Hasan Ibn al-Haytham (circa 965–1040)

Have you ever wondered how birds know when it is time to migrate, or how mammals know when it is time to hibernate? Decades of research has shown that these rhythms of behavior are not driven by animals simply responding to the environment but, instead, occur because of changes inside the animal. These internal changes are due to genes and physiological systems that help an organism to anticipate environmental changes. How the change in physiological systems is controlled has puzzled scientist for many years. When we look at the diverse range of physiological systems or functions what we see is that evolution has enabled us to keep our internal "state" remarkably *stable*. That does not mean that our internal state of physiological stability is always the same; it may vary predictably over time.

Predictable or anticipatory changes in seasonal physiology and behavior vary from well-defined periods of pregnancy to the daily fluctuation in leaf movements. There are evolutionary conserved genetic programs that orchestrate the regulated changes in physiological stability. Walter B Cannon is considered by many physiologists as the individual who formalized our understanding of internal stability in physiological systems and coined the term *homeostasis*. Since the 1920s, homeostasis has been the cornerstone for teaching physiology at multiple educational levels and research

On Rheostasis. Tyler John Stevenson, Oxford University Press. © Oxford University Press 2024.
DOI: 10.1093/oso/9780197665572.001.0001

programs. Open any university biology or medical textbook and the introductory chapters will devote a significant portion to the role of homeostasis in almost all physiological systems. Even during the early 20th century, it was recognized that homeostasis was a powerful model for understanding the inner workings of animals and plants, yet the limitations when applying this principle across all physiological systems have largely been ignored. Toward the end of the 20th century, Nicholas Mrosovsky started to compile a list of physiological responses that did not adhere to homeostatic principles. The evidence indicated that another physiological response induced *regulated* changes in internal stability, and he published the evidence in *Rheostasis: the physiology of change* (1990). Many of the phenomena identified by Mrosovsky included long-term rhythms or physiological responses that had a well-defined cycle (e.g., daily, seasonal) or covered distinct periods of time (e.g., pregnancy, fever). Physiological processes that showed regular, predictable, and repeated cycles of change were described as *programmed rheostasis*. A second *reactive rheostasis* process was developed to characterize similar regulated changes in physiological stability that occurred over a defined length of time.

The overall aim of this book is to build on the physiological systems initially characterized by Nicholas Mrosovsky and to use genetic, molecular, cellular, and anatomical evidence to illustrate that the principles of rheostasis can be dissociated from those that are based on homeostasis. The book intends to highlight that rheostasis is an evolutionary ancient and innate response that acts to regulate most physiological systems, independent of common stimulus-response homeostatic mechanisms posited by Walter Cannon. The book begins by describing the evidence for long term changes in physiological stability observed in nature using clear and well-established examples in Chapter 1. Here, the focus is on examples of long-term regulated changes in diverse physiological systems across the animal kingdom. Chapter 2 defines the concept of rheostasis from the perspective of programmed and reactive regulated changes in physiology and how these build on homeostatic mechanisms. Chapters 3, 4, and 5 provide three detailed examples of how programmed rheostatic changes in the circadian control of daily rhythms, female reproductive cycles, and seasonal rhythms in biology can be dissociated from homeostatic changes in physiological stability. Chapter 3 shows that the endogenous, genetic programs of daily timekeeping provides a predictive change in most physiological systems. The circadian clock eloquently illustrates how some homeostatic mechanisms can change in a programed manner. Chapter 4 expands on the concept that the genetic make-up of circadian timekeeping is involved in long-term stability in physiology and emphasizes that female reproductive cycles consist of central pattern generators that govern predictable periods. Chapter 5 provides a further example of programed rheostasis whereby seasonal rhythms in energy stability show that yearly changes in energy rheostasis serve to facilitate survival. Chapter 6 turns to examples of reactive rheostasis to illustrate how homeostatic mechanisms remain constant despite a temporary and regulated change in physiological stability. Here evidence collected from the control of an evolutionary conserved defense mechanism, fever, is used to demonstrate that the regulated change in core body temperature is governed by reactive rheostatic processes. Next, Chapter 7 provides a second example of reactive rheostasis and reviews the regulated changes during mammalian gestation and avian incubation. Chapter 8 introduces the concept of *allostasis*, a process in which physiological stability is maintained through change. In this chapter, the stress response is used to propose that multiple physiological systems of stability (i.e., homeostasis, rheostasis, and allostasis) exist and form levels of hierarchical organization. Chapter 9 outlines how each level of physiological stability adheres to different mathematical modeling to support the argument that homeostasis, rheostasis, and allostasis are governed by distinct genes, cells, and brain regions. Finally, Chapter 10 discusses the impact of disruption to rheostasis and how conflicts in rheostasis develop, with a focus on implications for pathological conditions, such as chronic illness and disease.

The book is intended for students, clinicians and veterinarians, and the public. It is hoped that if you are new to this area, and the information excites you, that you will exercise patience with the physiological concepts described in each chapter. If you are a seasoned expert, some concepts may lack the scientific rigor seen in scientific journals to reach the intended balance. Where crucial evidence is lacking, the keen researcher or junior scientist is encouraged to conduct the pressing experiments. I do not claim to have solved all the problems in the control of physiological stability, but instead attempt to provide clear and well-argued reasoning to stimulate the reader, especially advanced undergraduate and graduate students, clinicians, and the motivated and courageous non-specialists.

My research on long-term stability in physiological systems has benefited from generous funding from multiple organizations. During the time spent writing this book, the Leverhulme Trust has kindly provided funding for laboratory study of the genetic basis of seasonal rheostasis for energy balance and reproductive physiology. Over the course of my research career, several other agencies have funded many of the ideas presented in this book, including the Natural Sciences and Engineering Research Council in Canada, the National Institute of Health in the United States, and the Biotechnology and Biological Sciences Research Council in the United Kingdom. Finally, I am grateful for the scientific community and financial support provided by the British Society for Neuroendocrinology. I acknowledge that figures in the book were created using BioRender, including Figures 1.3, 2.2, 3.2, 3.3, 4.4, 5.1, 5.2, 5.6, 6.2, 6.4, 7.2, 7.3, 7.5, 8.1, and 8.2.

Several colleagues provided valuable feedback on earlier chapter versions. I thank Neil Evans, Fran Ebling, Dan Haydon, Brian Prendergast, and Scott MacDougall-Shackleton for their expert input. I also thank Elisabetta Tolla, Gaurav Majumdar, Christopher Marshall, Adam Liddle, Irem Denizli, Andrew Stevenson, and Calum Stewart for valuable discussions. Finally, I thank my wife Andrea and daughters Lilah and Charlotte for their patience and understanding during the conceptualization and writing of the book.

1

Long-term physiological stability in nature

We have so much time and so little to do. Strike that, reverse it.
Roald Dahl (1916–1990),
Charlie and the great glass elevator

While growing up in rural southwestern Ontario, it was remarkable to observe the diverse variation in yearly strategies that organisms used to survive or take advantage of annual changes in a temperate-zone climate. The annual migration of tundra swans (*Cygnus columbianus*) occurred with an extraordinary level of precision (i.e., migrating through on 10th March ± 4 days) and reflected the well-timed train schedules of continental Europe. Other relatively stationary animals such as the Eastern cottontail rabbit (*Sylvilagus floridanus*) have adapted to the local annual change in climate by shifting their reproductive activity to coincide with the dramatic increase in food resources during the summer. Because of the stark changes between summer and winter in Canada, two seasons have been facetiously characterized. A winter phase, in which inhabitants either migrate south to escape or bunker down in a dwelling in order to survive the cold harsh conditions. Then comes the construction phase, the time for building and infrastructure repairs or for harvesting crops in preparation for the next winter.

These examples are a very small snippet of the large range of seasonal behaviors that plants and animals use, from insects to frogs and birds to mammals, which have evolved as *regulated* annual changes to ensure the survival of the species. In addition to these long-term changes, relatively short-term daily changes in an organism's behavior are evident. The daily sleep-wake cycle is a robust, repeatable, and functionally significant behavior that is regulated by a well-described internal "clock." There are also other cyclical physiological and behavioral changes that occur with a regular periodicity that extend beyond a daily but less than an annual period. The reproductive cycle of most female rodents recurs with

On Rheostasis. Tyler John Stevenson, Oxford University Press. © Oxford University Press 2024.
DOI: 10.1093/oso/9780197665572.003.0001

a predictable 4–5-day cycle. Similar reproductive cycles occur in several primates, including humans, which approximates the lunar cycle (i.e., 29.5 days) or semilunar cycle (i.e., 14.8 days) periodicities. Indeed, the intertidal zone, or the geographical region that is above water at low tide and underwater at high tide, has several plants and (in)vertebrate organisms that maintain physiological and behavioral cycles that coincide with the lunar cycle (12.5hr). For example, a sea louse (*Eurydice pulchra*) off the coast of Wales shows consistent swimming activity cycles that are temporally fixed to the tide.

In some very rare cases, predictable long-term physiological and behavioral cycles have evolved to span extreme periods that are more than 1 year. A classic example is the periodical cicadas in the eastern region of North America. The cicada (*Magicicada septendecim*) resides in the ground as juveniles and feed on plant roots for 13 or 17 years then emerge in late April or early May as adults all at once in the same year. Late in the nineteenth century, the US entomologist Charles Lester Marlatt (1853–1954) characterized 30 different broods of cicadas in North America: 17 had 17-year life cycles and the other 13 had 13-year life cycles. As my graduate tenure at The Johns Hopkins University in Baltimore, Maryland, was coming to an end, a large emergence of the Great Southern Brood (XIV) occurred, and the next emergence of brood was predicted to occur in 2024. Examples such as the life cycle of the cicada provide clear evidence that relatively long-term internal programs exist to time essential life history transitions and serve to optimize an organism's fitness.[1]

Geophysical factors are at play, such as the 23.4° axial tilt (or obliquity) of the Earth, which is the primary reason for the seasonal changes in climatic conditions that develop because of the annual cycle around the sun. The tilt produces strong seasonal variation in environmental cues, such as the annual change in day length, temperature, and precipitation with large predictive patterns in habitats from the latitudinal range of 23.5° to the (ant) arctic poles (Figure 1.1). This does mean that geographical regions in the equatorial or tropical zone do not experience seasonal variation in environment conditions. Rainfall is one climatic cue that readily shows predictable cycles in the tropics, and most plants and animals use annual rainfall as an external cue to time seasonal cycles in physiology (e.g., breeding). Some animals that inhabit the equatorial zone are also capable of detecting very small changes in day lengths across the year. When Micaela Hau was working at the Smithsonian Institute, she discovered that the Spotted antbird (*Hylophylax n. naevioides*), which lives around 9°N in the Republic of

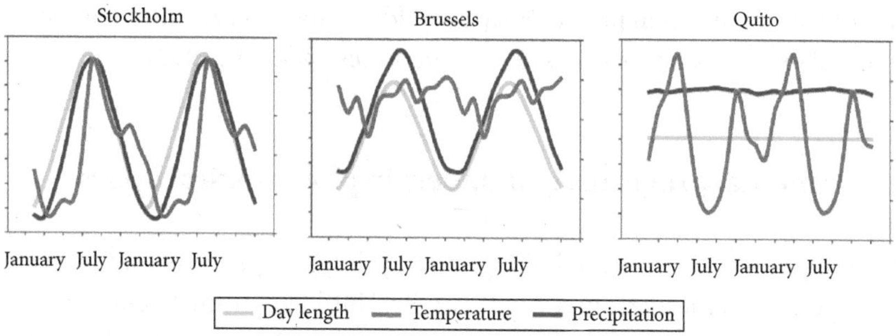

Figure 1.1 Long-term geophysical variation in the environment.
Seasonal changes in the local climate are primarily driven by annual oscillations in day length, temperature, and precipitation. Stockholm experiences large variation in all three climate cues, which are synchronized across the year, providing an extremely predictable cycle. Brussels exhibits synchrony between day length and temperature; yet, the seasonal change in precipitation is much less pronounced and highly variable. Quito lacks substantial changes in annual day length and temperature, yet exhibits marked bimodal phases in precipitation. These patterns demonstrate that both temperate and tropical regions experience seasonal environments and the annual change in climate cues (e.g., day length) can be used to determine the time or phase of the year.
Source: Data adapted from Stevenson et al., 2015.

Panama, responds to annual changes in day length of little as 17 minutes (Hau et al., 1998).

Three other periodic movements in the Earth's orbit contribute to regular, extreme long-term cycles. The Serbian geophysicist and astronomer Milutin Milanoković (1879–1958) described three cycles due to (1) cycles in the shape of the Earth's orbit (changes in eccentricity), (2) cycles in the Earth's wobble (axial precession), and (3) cycles in the tilt of Earth's axis (changes in obliquity). The orbits of the Earth around the sun are not constant and have been charted to move from a near perfect circle to a slightly elliptical shape with the cycle between shapes taking approximately 100,000 years. The Earth's rotation also has a "wobble" around the axis that has a regular 26,000-year cycle. And lastly, the change in the Earth's obliquity occurs over a 41,000-year cycle and is due to the change in the tilt from 22.1° to 24.5°. Although Milankovitch cycles will impact organismal lineages across multiple generations, the lifespan of a single generation (or a few generations) only spans ecological time (i.e., tens to hundreds of years). Therefore, animals are not under natural selection pressures that would facilitate the evolution of physiological processes over such extreme time scales (Bennett, 1990).

Internal programs that underpin physiological cycles

For the purposes of this book, long-term cycles include those that oscillate for a period that lasts from a day to a year. Until late in the twentieth century, the consensus was that plants and animals simply react to local changes in environmental conditions. This is understandable as there are ample examples in which an event such as the spring warming of the marine environment drives an explosion in the algae populations along the eastern coast of North America and the shores of France, Ireland, and the United Kingdom. Even today, the rapid warming of the Atlantic and Pacific oceans, due to climate change, is facilitating a northward expansion of harmful algal blooms (Gobler et al., 2017). Yet, in laboratory conditions at the Woods Hole Oceanographic Institution, Donald Anderson and Bruce Keafer (1987) discovered that algal blooms could occur independent of any environmental change. A specific type of marine algae, the dinoflagellate (*Alexandrium tamarense*) which can play a key role in algal blooms, has a wide distribution, and is readily observed off the coast of North America and Western Europe. The annual life cycle of this single-celled organism consists of motile vegetative states and resting cysts. During the winter, coastal cysts are dormant for 2–6 months, during which time they cannot germinate into the motile cells that cause massive algae blooms. Anderson and Keafer found that *A. tamarense* cysts that were maintained for several months at 2°C in constant darkness would germinate into motile cells with a near 1-year period and that this pattern persisted in constant laboratory conditions for 2 years! Anderson and Keafer proposed, therefore, that an internal program for cysts to germinate into motile cells was critical for the organism to anticipate warming waters in the subsequent year as opposed to reacting to warming waters and thereby facilitated reproduction during the optimal water temperatures. These findings are remarkable, as they indicate that a long-term internal timer exists even in a single-celled organism and that such a biological timing system has been a component of the genetic makeup of this species for several million years (Lincoln, 2019).

The benefits of long-term programs in physiology are many, though the primary advantage is that internal programs provide for a predictive physiological change so that an organism will *anticipate* a future event. This characteristic of internal timing mechanisms provides for a pre-emptive change and allows an organism to adapt to an upcoming environment, which serves to increase the probability of survival. One long-standing question has been, "how are these physiological programs controlled?" One explanation is that plants and animals simply respond to climate cues, such as temperature. The environment warms up and the organisms adapts to the change. As we have discussed above many organisms, such as *Alexandrium* and many insect species reproduce and multiply at an exponential rate in warm environments. However, a long-term "clock" will permit an anticipatory change in the absence of any change in the environment. These types of responses are referred to as *endogenous* change, a form of internal timer that ensures animals' physiological systems and behavior prepare for different environmental conditions.

Similar long-term internal timers have been observed in organisms that live for many years. Evidence of one such timer is provided by species which undergo hibernation. Both the Canadian grizzly bear and the polar bear hibernate for 5–7 months each year during the winter when food supplies are scare, and then each spring, they emerge from their dormant state in preparation for the spring feasts. Similarly, most rodents living in temperate zones display seasonal changes in multiple physiological systems to cope with winter environments. There is a diverse range of adaptations, including an inhibition of reproduction and loss of body mass and fat stores during the winter months. Rodents invest in reproductive physiology and sexual behavior during the spring and summer conditions when the nutrients (i.e., seeds) available for offspring are abundant. Then during the fall, rodents will cease reproductive investment and store energy resources to survive the winter.

In the laboratory, the Siberian hamster has received considerable attention, due to the well-described effects that day length (also referred to as photoperiod) has on the timing of seasonal cycles in physiology (Lewis and Ebling, 2018). In their native habitat, hamsters can live for many years and experience some of the harshest environmental conditions. During spring and summer conditions, hamsters engage in high levels of food intake, gain an impressive amount of body mass, and produce several litters of offspring. Decreasing day length in the autumn, however, results in the termination of breeding, and hamsters lose between 30% and 50% of their body mass over the span of a couple weeks. The hamsters have taken the short days as a predictive cue that cold temperatures, snow, and a lack of food resources are on the way. In other words, the decreasing day lengths after the autumnal equinox indicate that "winter is coming." To survive the winter, hamsters will burrow, creating a microclimate, and then, like bears, engage in a form of hibernation called torpor. The key difference between hibernation and torpor is that torpor occurs daily and is characterized by short phases of reduced activity, shivering, and decreased body temperature. The 5–6-month dormant winter state can easily be observed in the laboratory or even in household pets. The fur of Siberian hamsters housed during short days will turn white and the animals will spend a significant proportion of their time in bouts of "shivering" in their burrows, a clear sign of the torpor state. Even in the presence of warm temperatures (e.g., 22°C) and ample food, hamsters will decrease food intake by over 25% and build nests to support thermoregulation. After 5–6 months in the short "winter-like" light conditions, the hamsters will spontaneously change their fur back to an agouti color, gain substantial body mass, and no longer show torpor. This spontaneous change in physiology and behavior is the anticipatory change that prepares hamsters for the spring climate. The long-term endogenous, programmed change in hamster body mass has received considerable attention from a biomedical perspective, because the equivalent body mass change in humans would be a 200-pound individual losing over 60 pounds in a span of 3 months. It is likely not surprising that hamsters have provided an excellent animal model for understanding the genetic, molecular, and physiological basis of body mass regulation in humans (Stevenson et al., 2016).

Migratory behavior in wild birds provides a visual cycle in nature and is easily quantifiable. Seminal work was conducted in this field by Eberhard Gwinner (1938–2004), a German ornithologist with a particular interest in the physiological control of migratory behavior in songbirds. In his book *Circannual clocks*, Gwinner provides a clear overview of seasonal cycles in reproduction, moult, migration, and body mass of many songbird species. For example, the European stonechat (*Saxicola rubicola*) is a migratory species that spends the breeding period in western Europe and then migrates south to Africa for the winter. In February, the stonechats arrive in Europe and in September–October they migrate back south for the winter. In his laboratory, Gwinner studied the European stonechat's equatorial relative, the African stonechat (*Saxicola torquatus*), and discovered that these birds will display an increased level of restlessness activity during the "night," a behavior that reflects migratory behavior and is called zugunruhe. By keeping these birds in a constant day-night schedule (i.e., 12-hr light, 12-hr darkness) for 12 years, Gwinner tracked reproductive physiology, moult, and zugunruhe cycles all of which exhibited annual periodicity of approximately 10.5 months. Like the spontaneous increase in body mass in the Siberian hamster, zugunruhe behavior in songbirds is driven by an internal program that anticipates a future event. This work has been expanded by Vinod Kumar at the University of Delhi in India. Using the migratory Black-headed bunting (*Emberiza melanocephala*), Kumar and his group discovered that a change in laboratory light conditions can induce a seasonal program that reliably produces zugunruhe behavior. Buntings are highly photoperiodic and remain in a non-breeding, non-migratory state when housed in a short day (i.e., 8-hr light) condition. Transfer of birds to a stimulatory long day (i.e., 16-hr light) will result in reproductive development and body mass gain. After two weeks of long-day stimulation, birds will increase zugunruhe behavior during the dark phase. Finding analogous migratory cycles in mammals is challenging, yet the annual migration of reindeer (*Rangifer tarandus*) in the arctic or Blue wildebeests (*Connochaetes taurinus*) in Africa are two excellent examples.

When Gerald Lincoln (1945–2020) was at the University of Edinburgh, he studied a wide range of ungulate species and sought to identify the physiological and molecular basis of seasonal cycles. His research primarily used either the red deer (*Cervus elaphus*) population on the Isle of Rhum off the west coast of Scotland or various sheep species (*Ovis aries*) both in the wild and in the laboratory. The advantage of using red deer is the high level of seasonal cycling that coincides across the population. The breeding season, or rut, is associated with the "weaponization" of antlers and increased territorial behavior such as "roaring." Lincoln identified that hormones are critical for the seasonal cycle in antler morphology as well as sexual and aggressive behavior. For example, gonadal steroid testosterone production was dependent on the month of the year, and implants of testosterone increased antler growth, copulatory behavior, and territoriality. His research was essential in demonstrating that one key marker of the seasonal phase was the hormone prolactin and was a clear "read-out" of the brain signaling time of year (Lincoln et al., 2006). In his paper aptly titled "A brief history of circannual timing," Gerald elegantly describes how annual cycles are present across the animal kingdom and can occur in *synchrony* or *asynchrony* within a population. Most seasonal cycles observed in nature show synchrony, meaning that most of the population is in the same seasonal phase. A good example of seasonal synchrony is the annual timing of lambing. In the early days of courting my to-be wife, her family maintained a flock of sheep at their home in southwestern Ontario. With a very high level of predictability, the births of lambs would begin on what seemed the coldest night in February. Braving the −20°C temperature during the night period, the family would assist in the birth of lambs. In Scotland, where I now live, lambs are born during a two-week window in the month of March. The timing of lambing seasons is critical for the population. Lambs initially gain necessary nutrients from suckling their mothers. After a developmental period of several weeks, they gain independence (or perhaps the mothers regain their independence) and start to graze on rapidly growing grass in the Scottish moors. Yet not all animal populations display a seasonal event that is observed simultaneously across the population. A classic example is the breeding cycle of socially dominant male African elephants (*Loxodonta* sp.). African elephants living in equatorial areas experience only small changes in environmental cues. Yet, elephants show a remarkable degree of predictable timing in annual breeding cycles, known as musth. While musth can be observed across the entire year, it is entirely dependent, however, on an individual African elephant's reproductive cycle. The observation that dominant males engage in musth at different times of the year might suggest that these animals are *aseasonal*, or lack an annual cycle in reproduction. However, an astute naturalist monitored the individual cycles of elephants and discovered that each bull showed a reliable annual musth at the same time of year (Poole, 1987). Some male elephants would reliably show musth during the months September–December, whereas other members of the population would reliably musth during the months January–April. The different annual timing of individual dominant males shows that within a species there can be asynchrony in physiological and behavioral patterns. Subordinate males do not show musth cycles. Only when dominant bulls die can a subordinate take over the preferred breeding phase. Overall, the population shows a level of *asynchrony* in seasonal reproduction, but the individual bulls display a remarkably high level of seasonal cycling.

Most of the evidence for internal programs in physiology is derived from studies that investigate the endogenously generated daily cycles. Researchers and amateur naturalists have long known that plants, insects, and "higher" animals exhibit clear daily changes in physiology and behavior. For example, in 1736 Johannis De Gorter demonstrated that the human body temperature in a typical environment shows a reliable cyclical daily change from approximately 36°C to 38°C. It is now well established that most internal processes do not simply react to the daily changes in the environment, and instead there is an internal program that provides all organisms with a timer that approximates 24 hours. This internal daily timer operates even in the absence of external cues, and most physiological systems that exhibit daily cycles are driven by an internal program referred to as a *circadian* clock. The term circadian is derived from the Latin *circa* meaning "around" and *diem* meaning "day." The most common experiments to confirm the presence of an internal daily clock use animals that are maintained in a constant day length (i.e., 12 hr light, 12 hr darkness), with food constantly available, and in a temperature maintained around 22°C. The animals are then exposed to a simple manipulation in the environment—a switch to constant darkness, or constant light. This simple procedure places an organism (e.g., a mouse) in an environment that is devoid of environmental cues (i.e., light), so doing away with anything that could provide time of day information. In the absence of any environmental cue, only under the influence of an internal timekeeper will the animal be able to maintain a near 24-hour cycle. Data from such an experiment is shown in Figure 1.2 where the test subjects were mice, which are nocturnal animals, and can readily adapt to constant darkness. In a 12-hour light, 12-hour dark daily schedule, the nocturnal mouse shows high locomotor activity during the dark phase and rest/sleep during the light phase. So even though most animals on the planet are subjected to daily variation in environmental cues (e.g., sunrise, sunset), these cues act to *entrain* or maintain rather than drive daily cycles.

It is now established that multiple timescales are also represented internally and continue to cycle over set periods in the absence of environmental cues. The studies described above—including algae blooms, hibernation and torpor, and zugunruhe in laboratory conditions—established the existence of internal programs in physiology that occur over annual timescales. Similar Latin terminology is used to describe these long-term internal timers, including *circaquadridian* (4–5 days), *circalunar* (28 days), and *circannual* (365 days). The distinction between generic terms such as "seasonal" and Latin terms such as circannual is important. These terms help to clearly articulate the difference when discussing a cycle observed in nature and entrained by the environment, such as the "seasonal" tundra swan migration or "yearly" *Alexandrium* blooms in the north Atlantic Ocean, from internal physiological timers like "circannual" clocks identified in the absence of environmental cues, such as zugunruhe in Gwinner's stonechats.

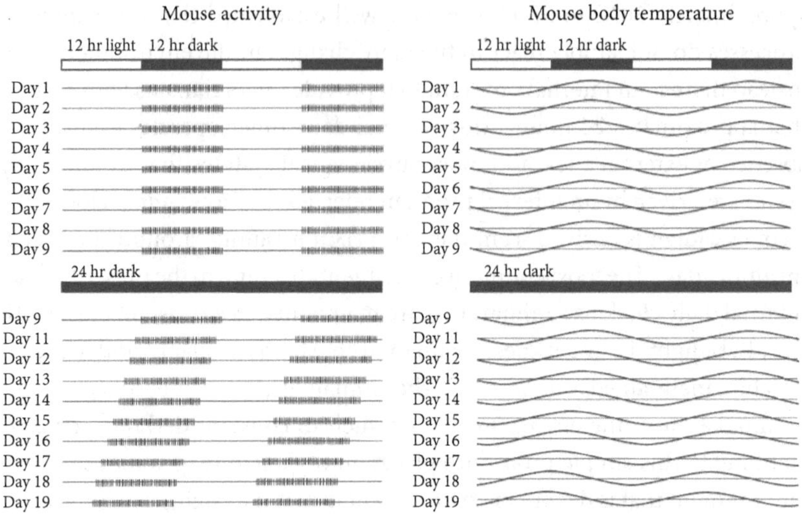

Figure 1.2 Behavioral visualization of the endogenous circadian clock. Mice models have been valuable for our understanding of the behavioral and genetic control of the circadian clock. A simple test to show the endogenous rhythms is to maintain mice in a 12-hour light, 12-hour dark cycle (indicated by white and black boxes). Mice are nocturnal and are generally inactive during the light phase and show higher locomotor activity during the dark phase (indicated by hatch marks). The data are provided in a "double-plot," in which two days are shown side-by-side to highlight the behavior across multiple cycles. The locomotor behavior of mice is reliable, robust, and consistent in 12:12 light:dark cycles. To observe the internal clock, mice are moved to a constant darkness environment, which lacks any daily signals (on day 9). Exposure to constant darkness maintains the near 24-hour cycle in activity, but prolonged exposure leads to a shift to earlier "daily" activity. The slight advance in activity from day 11 to day 20 is referred to as the "free-running" rhythm. The activity is maintained near 24 hours (approximately 23.5 hours) due to the internal circadian timer. Most rhythms are controlled by the circadian clock, such as body temperature (right column).

Long-term physiological timers in humans

Conducting similar experiments in humans is especially difficult. The availability of an environment with no external cues (e.g., daylight, sound) and human motivation for being isolated for prolonged periods of time are two key challenges that have limited the capability to advance our scientific knowledge on daily internal clocks in humans. During the 1960s, however, Michel Siffre, a French geologist and avid speleologist (someone who studies caves) became interested in the impact of living in caves where he was unaware of the time of day. He kept track of his eating and sleep-wake schedules, body mass, and body temperature and reported this information to team members camped outside the entrance to the cave. After several months, Siffre's physiological and behavioral data showed clear predictable cycles that approximated a 24.5-hour cycle. The inception, preparation, and descent to the Scarasson Cavern used for the cave studies is described in his book *Beyond time*. Siffre's work has been instrumental in demonstrating that humans have an internal clock that provides individuals with daily time information. The evidence for circadian cycles and the genetic underpinning of the circadian "clock" is extensive. Those particularly interested in the circadian control of life on earth are encouraged to read an excellent book titled *Rhythms of life* by Russell G Foster and Leon Kreitzman.

It is likely that many physiological cycles are underappreciated due to the relative lack of attention that has been given to individual human measurements over the long term. Indeed, it is often said that "humans do not have long-term programs in physiology," or "humans are not seasonal," but that view continues to be challenged. It is difficult to study individual humans for extended periods of time and, crucially, to not lose participants due to drop out. Therefore, it is sometimes necessary to use other scientific methods to study seasonal cycles in human physiology, for example the examination of historical datasets. In the late 1980s, Till Roenneberg and Jurgen Aschoff used monthly birth rates from 166 countries and spanning 3000 years to demonstrate that human reproduction shows a remarkably stable seasonal pattern in conceptions, with the peak occurring close to the spring equinox. This finding was further strengthened by the demonstration that there was a strong hemisphere effect, with regions in the Southern hemisphere found to have a 6-month shift in peak conceptions with greater levels in the autumn (Roenneberg and Aschoff, 1990). Additional, similar studies have yielded extensive data to support the idea that humans exhibit seasonal cycles in a range of physiological, immunological, and behavioral traits. A full overview has been provided by Russell Foster and Till Roenneberg in their paper "Human responses to the geophysical daily, annual and lunar cycles" (2008). A consequence of this human seasonality that has gained significant attention is that disease incidence can also be seasonal. One of the best characterized seasonal diseases in humans is multiple sclerosis. Both the onset and prevalence (i.e., start and number) of symptoms increase from March to May in the Northern hemisphere and October to December in the Southern hemisphere. Even many childhood diseases show reliable seasonal incidence, which cannot be accounted for by environmental factors such as term-time (Bakker et al., 2016). Moreover, the month of a person's birth also appears to affect prevalence for psychiatric disorders, neurological illness, or other general pathologies. For example, the prevalence of Hodgkin disease is higher if an individual was born in February–April, eating disorders are higher in individuals born between February and May, and Parkinson's disease is higher in individuals born in April through May (Foster and Roenneberg, 2008).

One long-standing conundrum has been whether patterns in human seasonal physiology are driven by internal factors such as hormones and/or biochemical reactions. One researcher that sought to resolve this issue was Franz Halberg (1919–2013) at the University of Minnesota who characterized the seasonal cycles in multiple physiological systems. His work revealed that prolactin levels in women have a clear seasonal cycle with higher levels during the winter periods compared to spring and summer (Halberg et al., 1983). Coincidentally, prolactin levels appear to be highest at the time Roenneberg and Aschoff's data indicated that births also increase. Halberg's work also highlighted seasonal patterns in energy balance with peak carbohydrate and caloric intake occurring in the late summer, which preceded maximal body mass in September and October. Interestingly, the peak in metabolic rate occurred in January, which makes this an ideal time to lose weight!

In the later years of the twentieth century and into the twenty-first century, there has been a significant increase in the availability of large datasets gathered from the public, such as the UK Biobank derived from the records of over 500,000 volunteer participants and intended to improve health and well-being through the prevention, diagnosis, and treatment of a wide range of acute and chronic diseases. Researchers and academics across the globe can investigate the data to uncover patterns or markers for specific life-threatening illnesses, such as cancer, stroke, or dementia. Analyses of data relative to birth months available in the UK Biobank has already revealed a large seasonal pattern, with an individual born from June to August having on average a higher birth mass and later pubertal development compared to individuals born at other times of the year. Similar databases have been established in other countries and use an international pool of volunteers. At the Institut fur Diabetesforschung in Munich, Germany, a massive study called the BABYDIET has been developed by Professor Anette-Gabriele Ziegler to investigate whether early intervention using a delayed introduction of dietary gluten could reduce the incidence of pancreatic islet autoimmunity in high-risk children (aged up to 3 months) of patients with type 1 diabetes. The database includes a range of samples of blood cells (i.e., mononuclear cells), fat tissue, circulating hormones, and immune markers. Using the dataset, collaborating British and German scientists discovered that white blood cells and fat tissue exhibit a large seasonal pattern of expression for hundreds of gene (Dopico et al., 2015). Again, as with the human birth data described above, this dataset showed an inverted hemisphere shift, which reinforces that it reflects a seasonal cycle. Seasonal changes in markers of immune function, the interleukin-6 receptor and C-reactive protein, were also noted and their presence peaked during winter periods in the Northern hemisphere. The combination of long-term historical records and modern day international collaborative datasets provides strong evidence that humans have seasonal cycles in physiology and immune function that directly impact on individual behaviors.

How are long-term changes in physiology controlled in animals?

So, evidence exists for naturally occurring long-term cycles across the animal kingdom. These cycles have internal timing mechanisms and can show population synchrony or asynchrony. Given that these patterns are present from single-celled organisms to complex animals, one would hypothesize that an evolutionary conserved timing system must exist. In animals, the hypothalamus is the key region of the brain that links the nervous and endocrine systems and is critical for the internal representation of time. The hypothalamus is located at the base of the brain almost at the center of the human skull. The structure is relatively consistent in fish, frogs, birds, and mammals, and consists of several highly inter-connected nuclei, functionally linked neural cell groups, that are critical for a range of physiological processes and serve to maintain physiological stability or homeostatic balance. Several physiological processes are *innately* controlled by the hypothalamus, and these include body temperature, water balance, and sleep cycles (Figure 1.3). As the hypothalamus is essential for many physiological processes that maintain a healthy individual, damage to the entire hypothalamus will result in death.

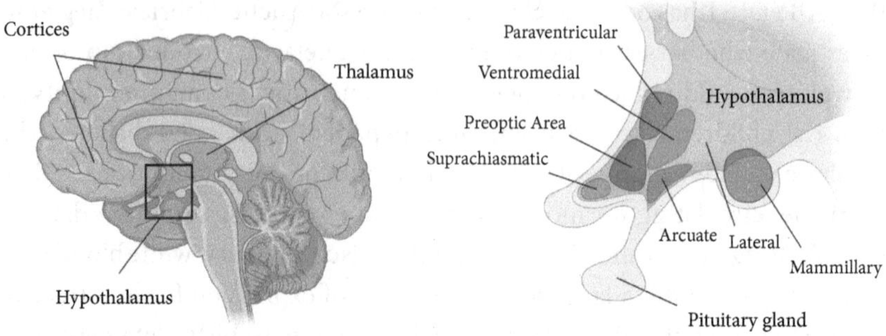

Figure 1.3 Representative neural substrates that regulate physiological stability. The hypothalamus is a cluster of nuclei that sits in the base of the brain. Several "extra hypothalamic" brain regions connect with the hypothalamus and convey information regarding emotional states (e.g., amygdala), environmental cues (e.g., thalamus), and contextual information (i.e., cortex/pallium). The hypothalamus is the primary relay center that integrates information from multiple sources and communicates with peripheral tissues using hormones released by the pituitary gland. The suprachiasmatic nucleus (SCN) receives light information from the retina in the eyes and is critical for the generation and maintenance of circadian rhythms in mammalian physiology, immune function, and behavior. The preoptic area (POA) is adjacent to the SCN and is necessary for the regulation of reproduction and body temperature. The arcuate nucleus (Arc) and ventromedial nucleus (VMN) are involved in timing female reproductive physiology and behavior. The Arc, VMN, and lateral hypothalamus act together to regulate energy homeostasis by signaling satiety and hunger. The paraventricular nucleus (PVN) contains neurons that are essential for pregnancy, birth, and the stress response.

However, during the mid-twentieth century, researchers started to conduct experiments that selectively damaged individual nuclei in the hypothalamus to identify the function of specific regions.

In the 1960s Curt Richter (1894–1988) working at Johns Hopkins University conducted a series of experiments that induced small areas of damage in the brains of rats to identify where the "center" for the control of various physiological cycles was located. Richter identified that if the optic tract that conveyed information from the eyes to the brain was severed, the daily cycle was abolished and rats ate, drank, and were active at all hours of the daily light and dark cycle (Richter, 1978). Only lesions to the suprachiasmatic nucleus (SCN) in the hypothalamus were capable of eliminating circadian regulation of drinking, locomotor activity, and daily hormone (i.e., corticosterone) rhythms (Stephan and Zucker, 1972; Moore and Eichler, 1972). From the work of Elizabeth Maywood and Michael Hastings at the Laboratory of Molecular Biology and the University of Cambridge, it is now known that within the SCN, a small population of approximately 1000 cells is critical for the integration of light cues that entrain daily cycles in physiology (Hastings et al., 2018).

Similar research to identify the regulatory centers that govern food intake and energy balance was conducted by Hetherington and Ranson. Using localized electric stimulation to induce lesions to the lateral hypothalamus, they found that following such damage rats showed aphagia and massive weight loss. Conversely, lesions that included both the ventromedial and arcuate nucleus in the hypothalamus produced hyperphagia and obesity (see Elmquist et al., 1999 for an excellent review). These findings led to the formulation of the "dual center" hypothesis of how the brain controls food intake and appetite. We now know that a series of peptide hormones produced in the arcuate nucleus are sensitive to circulating hormones that signal energy state, and either trigger or inhibit food intake. Evidence that both anabolic and catabolic neuronal phenotype cells are found in the arcuate nucleus refuted the dual center hypothesis. The anabolic and catabolic effects of lesions of hypothalamic structures ("centers") are just the net effect of depleting both anabolic and catabolic neuronal cells from those regions. The evidence for hypothalamic neuropeptide hormones that regulate energy balance and food intake (or any physiological process) has a long, and rather amusing, history.

In the 1950s there was significant debate on whether the hypothalamus produced hormones that controlled change in peripheral physiology. One of the strongest challengers to the hypothesis that the brain produced hormones was Roy Greep of the Harvard School of Dental Medicine. At the time, studies of hormonal control of peripheral physiology focused on the pituitary gland, a tissue located at the base of the brain that connected the hypothalamus to the peripheral circulatory system. Roy Greep's research was fundamental in establishing that the pituitary produced hormones that controlled a wide range of physiological process. There was no evidence that nerves from the brain (i.e., hypothalamus) passed directly into the anterior portion, or adenohypophysis, of the pituitary, which gave the impression that the neuropeptides did not communicate with the pituitary gland. It was not until the 1970s that a rather contentious competition between two scientists produced a series of publications that clearly showed that neurons in the hypothalamus produced peptide hormones that were released into a blood portal that connected the brain to the pituitary. The blood portal, or infundibular stalk, allowed hypothalamic hormones (i.e., neuropeptides) to move into the pituitary gland to control the release of other hormones. Roger Guillemin and Andrew Schally were two physiologists that discovered specific neuropeptide hormones in the hypothalamus and subsequently demonstrated that the release of neuropeptide hormones from neurons could regulate physiological processes by acting via hormones released from the pituitary gland. Their aims, sometimes in collaboration but mostly in competition, were to identify the factors in the hypothalamus that controlled physiological processes such as growth and reproduction. The collaborative and fiercely competitive interactions contributed to a long-standing rivalry that lasted over 20 years.[2] Obtaining hypothalamic material in the volumes required to isolate hormones was the major limitation as the time. So, they obtained massive amounts of hypothalamic tissue from abattoirs. The discovery of specific neuropeptide hormones in the hypothalamus that controlled energy balance, reproduction, body temperature, and an individual organism's stress response was a major advancement in the 1970s and has led to the development of multiple new drugs for contraception and the control of growth. In recognition for their work, Guillemin and Schally were awarded the 1977 Nobel Prize in Physiology or Medicine, which they shared with Rosalyn Sussman Yalow who developed radioimmunoassay, a method that was fundamental for the measurement of hormones circulating in the blood stream.

Now, several hypothalamic neuropeptide hormones have been characterized and shown to be involved in a diverse range of physiological processes either directly or indirectly via regulation of pituitary hormones (Table 1.1). Most of the hypothalamic peptide producing hormones are found in specific hypothalamic nuclei and project to the median eminence where the neurons release the hormones into the infundibular stalk. The released peptide hormone can then act within the anterior pituitary gland, which releases other hormones that enter the blood capillary system and act on peripheral tissues. A second group of neuropeptide hormones in the hypothalamus project directly into the posterior pituitary, which is in effect an extension of the hypothalamus and from where hormones are released directly into the circulatory system.

Prior to the identification of the role of the hypothalamus in the timing of physiological processes, it had long been assumed that the primary tissue

Table 1.1 Hypothalamic and pituitary hormones

	Anterior pituitary	
Hypothalamic hormone	**Pituitary hormone**	**Physiological system**
Corticotropin releasing hormone (CRH)	Adrenocorticotropin Hormone	Cortisol hormone, stress and sugar metabolism
Thyrotropin releasing hormone (TRH)	Thyrotropin-stimulating Hormone	Thyroid hormone, body metabolism, and body temperature
Gonadotropin releasing hormone (GnRH) (+ve)	Follicle-stimulating Hormone	Maturation of the follicles and the production of sperm
Gonadotropin inhibiting hormone (GnIH) (−ve)		
GnRH (+ve)	Luteinizing Hormone	Sex steroids (i.e., testosterone, estrogen)
GnIH (−ve)		
Growth hormone releasing hormone (GHRH)	Growth Hormone	Growth hormone and metabolism and skeletal growth
Vasoactive intestinal polypeptide (VIP) (+ve)	Prolactin	Reproduction, metabolism and immune function, and water balance
Prolactin inhibiting hormone (Dopamine) (−ve)		
	Posterior Pituitary	
Oxytocin	Direct release	Social bonding, contractions during birth
Vasopressin (aka anti-diuretic hormone)	Direct release	Water balance

Note: (+ve) indicates a stimulatory effect, (−ve) indicates an inhibitory effect

responsible was the pituitary gland. The identification of specific peptide hormones in the hypothalamus that regulated pituitary function provided key neuronal populations that could integrate the information required for the long-term regulation of such physiological processes. Several laboratories and decades of intense research has characterized how peptide hormones in the hypothalamus, now referred to as neurohormones, act to maintain a relatively constant physiological balance, or *stability*. Using a system of feedback loops, neurohormones can maintain a physiological set point that is optimal for a range of processes. For example, the preoptic area nucleus in the hypothalamus has temperature-sensitive neurons that act to maintain core body temperature around 37.5°C (Zhao et al., 2017). Upon stimulation of cells that express an inhibitory transmitter, GABA, there is a reduction in core body temperature (i.e., hypothermia). Conversely, the inhibition of GABA cells in the preoptic area will result in an increase in body temperature (i.e., hyperthermia).

In order to understand how the hypothalamus integrates environmental cues to fine tune long-term endogenous physiological cycles, it is important to select animals that are best suited for the research question under investigation. Laboratory mouse models (*Mus domesticus*) have been extremely beneficial for biomedical research due to their rapid generation times, the high genetic similarity between individuals, and the ease with which genetic modifications can be induced into the genome to reflect specific human pathologies. Unfortunately, mouse models have nearly lost seasonal cycles, due in part to more than several thousand generations of inbreeding conditions. This does not mean the endogenous component of circannual cycles has been lost, as mice still show a small pattern of seasonality in litters and number of offspring (Drickamer, 1990). Other animals that maintain large fluctuations in regulated physiological stability are most suitable for examining how the hypothalamus controls long-term cycles. Three species that have been informative in the study of long-term physiological cycles are hamsters, quail, and sheep (Figure 1.4). There is a wealth of scientific literature spanning over 50 years on the morphology, physiology, and neuroanatomy of these species. Several laboratories located in North America, Japan, India, and Europe have used these animals due to their significance for agricultural industry (i.e., sheep) and their ability to maintain reliable, repeatable, and robust responses to climate cues while housed in laboratory conditions. When possible, the wider evolutionary conserved nature of the hypothalamic control of long-term physiological stability will be developed by considering data generated from laboratory studies of non-mouse animal models. Quail show several fascinating behavioral and physiological responses to light and social cues. During the 1980s and 1990s at the University of Bristol, Sir Brian Follett discovered that a single 1-hour light pulse provided to a quail housed in short winter-like days (i.e., 8hr light), but 12 hours after the light came on, would stimulate long summer-like day reproductive development and breeding conditions. Despite the light pulse occurring when the birds expect the dark period to continue, it activated light receptors in the hypothalamus that trigger the gonadotropin releasing hormone neuropeptide to stimulate the gonads to grow and placed quail in

Figure 1.4 Laboratory models for studying long-term changes in physiology. Common biomedical animal models have been selected for rapid generation times, which has resulted in the loss of many long-term rhythms such as seasonal physiology. Three examples of animal models used to study how the hypothalamus controls regulated changes in physiology are hamsters, quails, and sheep. Siberian hamsters reliably show large-scale seasonal rhythms in multiple physiological, immunological, and behavioral processes. The Japanese quail has been the leading avian model of seasonal physiology due to its robust and rapid response to changes in day length. Sheep are an excellent livestock species and a valuable model system for examining long-term rhythms in a more natural environment.

Source: Hamster and sheep photographs taken by the author. Japanese quail photograph kindly provided by Sir Brian Follett and used with permission.

a breeding state (Foster et al., 1985). To date, all neuropeptides have been implicated in the short-term (e.g., minutes to hours) regulation of physiological processes. But how do these neuropeptides function over long-term cycles, and how would these neuropeptide hormones interact with other physiological systems to ensure optimal coordination across organs, tissues, and cells? To address these questions, a conceptual framework is required to understand how long-term physiological stability is governed by the hypothalamus.

Physiologically programmed and reactive internal timers

Nicholas Mrosovsky (1934–2015) was a major intellectual and scientific contributor to the understanding of long-term cycles in nature. In many ways one could argue that he pioneered the current philosophical approach, which is impeccably articulated in his seminal book *Rheostasis: the physiology of change*. Mrosovsky held a Professorship at the University of Toronto, Canada, where he was on the faculty of the Department of Physiology, Psychology and Zoology. His research interests were diverse, spanned multiple physiological and behavioral systems, and readily examined some measure of time, either circadian biology or other long-term cycles such as seasonal biology. He published extensively on the physiology of hibernation and, working with Sara Shettleworth (also a professor at the University of Toronto), showed the impact of temperature on hatching sex ratios in sea turtles. Their work is still essential for conservation efforts and an excellent example of how scientific data can best inform policy. In his book, Mrosovsky outlined two distinct forms of long-term, regulated changes in physiology that have clear adaptive value: programmed and reactive rheostasis. *Programmed rheostasis* includes the regulated changes to a physiological variable during the life cycle of an organism. Conversely, *reactive rheostasis* describes a transient physiological response that occurs in the presence/absence of an external stimulus. Examples that can help to remember the difference between the two types of rheostasis include seasonal body mass/temperature cycles in hibernators (i.e., programmed) versus pregnancy (i.e., reactive).

Why should anyone care about the underlying control of rheostatic systems? Over the past few years, the impact of disrupted daily rhythms has been shown to have a profound detrimental effect that leads to increased probability and severity of general pathologies, neurological disease, and immune dysfunction. There are multiple examples in which prolonged shift-work or repeated jetlag increases the likelihood of obesity and reduced immune responses to common illnesses (Wulff et al., 2010). What has received considerably less intense study is the impact of disruption in rheostatic mechanisms on animal health and wellness. Some evidence that can be used to illustrate the adverse effects of altered long-term cycles can be derived from research investigating the impact of climate change on seasonal timing. The advancement in timing of "spring growth" has disrupted seasonal timing across ecosystems, has been observed in marine, oceanic, and terrestrial environments, and impacts plants, insects, and animals (Stevenson et al., 2015). For several decades, the annual migration, egg laying, and physiological responses of European Great tits (*Parus major*) have been monitored by a group of researchers at the Netherlands Institute of Ecology (NIOO-KNAW). The division head Professor Marcel Visser and his colleagues have examined the impact of local changes in climate cues, such as temperature, on the laying dates of the first clutch of eggs by Great tits. From 1980 to 2020, the average spring temperature has gradually increased with wide ranging phenological effects on the native plant, tree, and animal species in Western Europe. Caterpillar larvae emergence has adapted to earlier springs and the timing of their growth has advanced to occur several weeks sooner as well. As the caterpillar larvae are key food sources for the Great tits, this has also driven them to lay their eggs at an earlier date. To adapt to the change in food resources, the timing of egg laying in Great tits has slowly adjusted to an earlier lay date (Reed et al., 2013). Barbara Helm, Head of Migration at the Swiss Ornithological Institute, has also documented that the increase in ambient spring temperature has driven a near 1-day per year advance in egg-laying dates of a long distance migratory pied flycatcher (*Ficedula hypoleuca*) (Helm et al., 2019). Over a 21-year period, a wild population of pied flycatchers were found to advance egg-laying date by 11 days. In another population kept in a laboratory "common-garden" experiment, in which birds experienced the same annual photoperiod and temperature changes, multiple reproductive measures including molting of feathers, migratory behavior (i.e., zugunruhe), and egg-laying advanced by 9 days. Like the large historical datasets that have shed light on the seasonal cycles in human reproduction, decades of monitoring the activity of discrete populations of animals are required to identify the impacts of environmental perturbation on long-term cycles in physiological processes.

This book aims to extend the evidence outlined by Mrosovsky and highlight the hypothalamic control of long-term physiological changes by programmed and reactive rheostasis. The overall aim is to develop a theoretical framework that can account for how different neuropeptides (e.g., genes) and physiological processes interact over the long term to produce highly regulated physiological processes that serve to maintain physiological stability. The data will be drawn from plant and animal kingdoms and highlight that programmed and reactive rheostasis is present and forms complex, multi-level interactions to control whole organismal physiology. In Chapter 2 we will distinguish between the well-characterized homeostatic and rheostatic control of the physiological system and emphasize that length of time is one predominant characteristic that differentiates between the two systems. Subsequent chapters will use programmed changes in energy balance and reproduction or reactive programs in immune function and pregnancy to develop a framework to clearly dissociate homeostatic and rheostatic processes at levels of analyses that include genetic, molecular, and neural pathways and behavior. The central argument proposed is that physiological systems form a hierarchical organization in which time and genetic components can be used to easily establish that rheostatic processes form an independent level of regulated changes in physiological stability.

Notes

1. My advisor during graduate school at Johns Hopkins University described a point in his childhood in which there was a massive abundance of cicada in the streets of his neighborhood, which provided a windfall for an aspiring entrepreneur. Although, he humorously stated that a business could not last the intervening 16 years.
2. The story of the intense competition to discover and characterize the peptide corticotrophin releasing factor (CRF) now known as corticotrophin-releasing hormone (CRH) is beautifully described in the book *The Nobel duel* by Nicholas Wade, published in 1981.

2

Programmed and reactive rheostasis

Now, here, you see, it takes all the running you can do, to keep in the same place.

Lewis Carroll, *Through the looking-glass, and what Alice found there*

In *Through the looking-glass* by Lewis Carroll, the antagonistic Red Queen and Alice are engaged in a race in which both characters are constantly running but remain in the same spot. All life on the planet constantly faces a state of environmental randomness or uncertainty, and yet maintains a level of physiological stability. The continued protection of stability is required to ensure organs, cells, and biochemical reactions can function in a constant, optimal state. In evolutionary biology, there are multiple examples in which animals must constantly adapt, to the environment or in response to other organisms, in order to survive and reproduce. Animal physiologists have long known that an individual's body seeks to maintain a constant internal state despite large-scale variation in the environment. One of the first topics covered in most physiology courses is the concept of *homeostasis*. The term homeostasis is derived from the Greek words for "same" and "steady," and reflects the fundamental property of cells, tissues, and whole organisms to maintain and regulate internal processes for the stability and constancy needed for an animal to function at optimal efficiencies. In this chapter, we will cover the foundation of homeostasis and provide a series of examples to illustrate the key components of the central and peripheral nervous systems that maintain relative short-term stability in physiological systems. However, it is important to recognize that the state and stability of an individual can be assessed and optimized on a moment-by-moment basis, over the course of minutes, hours, days, years, and even decades. The longer-term regulation of physiological stability poses different challenges. A series of other theoretical models are described that attempt to cover long-term regulation of

On Rheostasis. Tyler John Stevenson, Oxford University Press. © Oxford University Press 2024.
DOI: 10.1093/oso/9780197665572.003.0002

stability in physiological systems. A model of *rheostasis* will be described that accounts for how long-term regulated changes in physiological systems are coordinated and how the mechanisms can be differentiated from homeostatic stability. The overriding conjecture proposed is that homeostasis serves to maintain stability in the short-term (minutes to hours) and acts with rheostatic systems that engineer stability over the long-term (days to years) to enhance survival. In this context, homeostasis is a primary stasis and is an evolutionarily ancient system, which is intrinsic to an animal and occurs entirely independent of experience. Rheostasis has properties that are like homeostasis but exhibit a larger degree in variability across the animal kingdom and include a level of species-specific responses. Conversely, some physiological systems can maintain stability through a change in an animal's behavior. The ability of an individual to adapt their physiological processes, mostly through experience with an environmental stimulus (or stressor) is referred to as *allostasis*. Allostasis is observed across the animal kingdom, yet there is a wide range of individual variation that is attributable to physiological systems that are modified by developmental and lifespan experiences. So instead of a single mechanism that governs physiological stability, we have a hierarchical model in which levels of stasis can be differentiated by genetic, cellular, neuronal, and organismal-wide properties.

Characteristics and principles of physiological stability

One of the first formal characterizations of mechanisms that seek to gain control over the internal environment of an organism was done by the French physiologist Claude Bernard (1813–1878), who coined the phrase *milieu interieur* or "the stability of the internal environment" to describe the conditions needed for a free and independent life. The concept of the milieu interieur was expanded several years later by the US physiologist Walter B Cannon (1871–1945). While working in the Department of Physiology at Harvard Medical School, Walter's research interests were broad and spanned digestion and acute stress responses. The reader keenly interested in the history of science may find Cannon's book *The wisdom of the body* (1932) a rather enlightening source. His research on the maintenance of blood fluid matrix (e.g., salt, glucose) lead to the conclusions that homeostasis was based on 4 key principles: (1) homeostatic systems seek to maintain a state of constancy, (2) a steady-state of activity requires that any change is automatically met by factors that resist that change, (3) there are typically multiple cooperating factors that form a complex system to maintain constancy in any physiological variable, and (4) homeostasis is organized *innately* and functions autonomously.

The core principles of homeostatic systems are that they seek to maintain a constant physiological state around a *set point*. To achieve this, Cannon's characteristics of homeostatic systems included a sensor, an error detector, a controller, and effectors that coordinate a rapid response to regulate a physiological variable (Figure 2.1). For any vertebrate (and some invertebrates), the levels of a physiological variable are monitored and detected by sensors or receptors located in the central and/or peripheral nervous system. Any deviation between the set point and the level of a physiological variable is integrated by an error detector. To restore the stability in the physiological system and maintain the variable within the normal range, the error signal is conveyed to a controller that signals an effector to adjust the levels of the variable closer to the set point.

One example to illustrate a homeostatic system is the control of body temperature in response to exposure to cold or hot environments. On average, the mammalian core body temperature is tightly maintained around 37°C during the day in diurnal animals. Here core temperature should not be confused with shell (or surface) temperature which can be significantly different from the core, as the shell (e.g., skin) forms one of the effectors to both protect the core body temperature and allow heat loss. The hypothalamus, the controller of the homeostat, functions to maintain the core temperature for anatomical structures such as the brain and heart at 37°C with

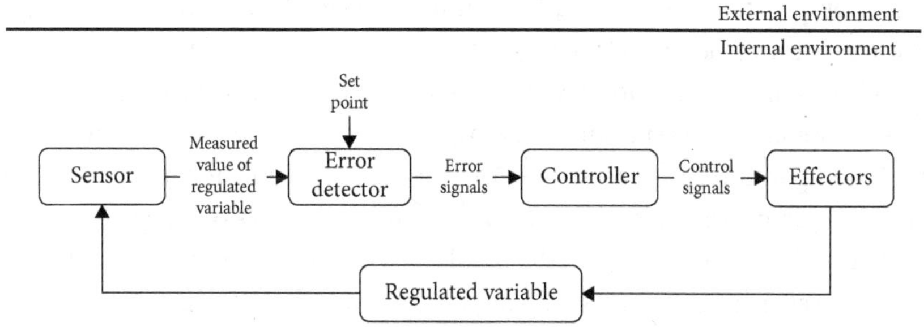

Figure 2.1 Schematic overview of homeostatic stability.
Cannon's four characteristics of a homeostatic system to maintain stability of a physiological variable include: a sensor, an error detector, a controller, and effector.

very little deviation (e.g., +/−0.1°C). Any significant decrease or increase in body temperature will have an adverse effect on normal cellular and tissue function, and if the hypothalamus cannot regain immediate control, the outcome will likely result in death (e.g., severe hypothermia). As outlined by Cannon, the homeostatic regulation of core body temperature is innate in the sense that it cannot be modified by individual experience. Sensors that respond to temperature include a range of temperature-dependent receptors located in the skin as well as receptors located inside the body (i.e., visceral) and the brain. These receptors detect changes in external and internal (i.e., core) body temperature and send the signals into the preoptic area nucleus in the hypothalamus. How the preoptic area detects both an increase and decrease in body temperature had long been a mystery. In 2017, several groups working at the Chinese Academy of Sciences discovered that a population of gamma-aminobutyric acid (GABA) cells in the preoptic area that when stimulated, could reduce body temperature. Inhibition of these cells resulted in increased body temperature and a fever-state in mice (Zhao et al., 2017). These findings suggested that the GABA cells in the preoptic area are continually responding to temperature signals from the skin, visceral, and brain sources and any error (i.e., deviation from 37°C) is detected. The error signal detected by these GABA cells in the preoptic area is then shared with adjacent nuclei, such as the dorsomedial hypothalamus. There, the dorsomedial hypothalamus integrates both warm- and cold-temperature information and acts as a controller to initiate physiological and behavioral changes to return to normal core body temperature. The effector component involves descending projections via the spinal cord to initiate the appropriate change to maintain stability in core body temperature. In cases when core body temperature increases, a human may engage behavioral changes and reduce clothing, or an animal might move into the shade, coupled with physiological changes as blood flow is increased to the hands/feet to increase heat loss to the environment and sweating is triggered. Conversely, when core body temperature decreases, effector responses will include an increase in clothing, or movement toward a heat source, and physiological responses such as reduced blood flow to the hands and feet to preserve body temperature, as well as triggering a shivering response to generate heat and activate non-shivering thermogenesis.

The homeostatic regulation of core body temperature is rapid and sensory, and effector mechanism are conveyed by the autonomic nervous system (Tan and Knight, 2018). In brief, the autonomic nervous system is a component of the peripheral nervous system and regulates a range of involuntary physiological processes. There are distinct divisions that include the sympathetic, parasympathetic, and an associated enteric pathway that provides sensory input and motor output to the central nervous system (i.e., hypothalamus). Dysfunction of the autonomic nervous system results in loss of function across multiple physiological systems and is life threatening. For homeostatic regulation of physiological processes, balance is predominantly immediate, and any deviation from the set point is rapidly re-established within minutes. For core body temperature regulation, sensory information is transmitted from skin and visceral tissues by neurons into the dorsal horn in the spinal cord. From here, ascending pathways project into the lateral parabrachial nuclei and then end in the preoptic area of the hypothalamus, a process that takes milliseconds. A similar pathway is involved in the effector control of thermoregulatory behavior and physiological responses. Projections from the preoptic area contact several regions in the midbrain, a "lower brain" region that acts as an intermediary between the brain and spinal cord. From the midbrain, there are specific projections into the interomediolateral column of the spinal cord that each control sweating, blood flow to the skin, and non-shivering thermogenesis. From the spinal cord, the sympathetic nervous system has preganglionic then post-ganglionic nerves that connect to sweat glands, smooth muscles, and brown adipose tissue to induce rapid changes in body temperature by heat loss or gain (Figure 2.2). Another pathway from the preoptic area is passed down into the ventral horn of the spinal cord and motor neurons connect with skeletal muscle to control shivering thermogenesis. For all the descending effector pathways there are 3–4 neuronal connections from the preoptic area to the target tissue, which again results in a rapid change (i.e., milliseconds) in core body temperature. Many animals, and especially humans, will also engage in behavioral thermoregulation. A series of cortical brain regions that control cognition in mammals, and pallial brain regions in non-mammalian species (e.g., reptiles), are important for the sensation and perception of whether the organism is "hot" or "cold." These higher-order brain regions drive behavioral thermoregulation that include moving into a warm/cold area, huddling with other members of the population, or changing one's outfit to better suit the environmental conditions.

The quick return to the core body temperature set point in reaction to an environmental change is one classic example of a "stimulus-response" form of homeostasis. Not all physiological systems follow this relatively simplistic

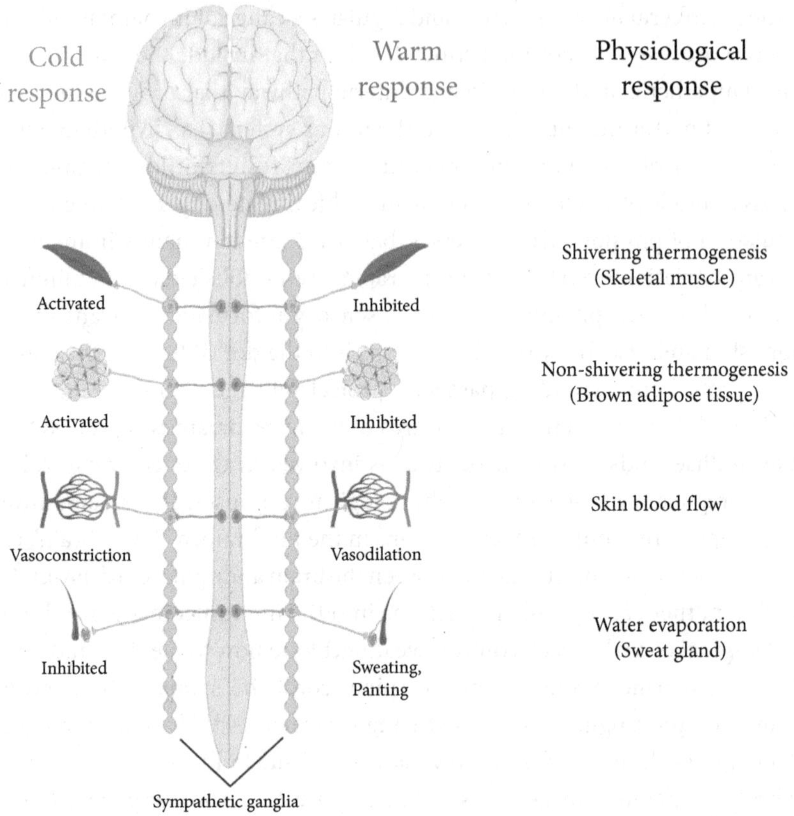

Figure 2.2 Schematic representation of sympathetic nervous system control of core body temperature.
Effector control of mammalian core body temperature includes sympathetic pre-ganglion connections from the spinal cord that connect to ganglion nerves located in the sympathetic ganglia. Ganglion nerves project to target tissues to activate physiological responses and rapidly maintain physiological stability. Cold environments activate skeletal muscle to induce shivering and adipose tissue to stimulate non-shivering thermogenesis. Blood vessels are constricted to prevent blood flow to the extremities and water evaporation from sweat glands is inhibited. The converse physiological response occurs in a warm environment.

model. Another example to highlight a fundamental property of homeostasis relies on the concept of negative feedback. The key difference in this example is that stability is maintained in the body without any input from environmental conditions. The concentration of several circulating hormones

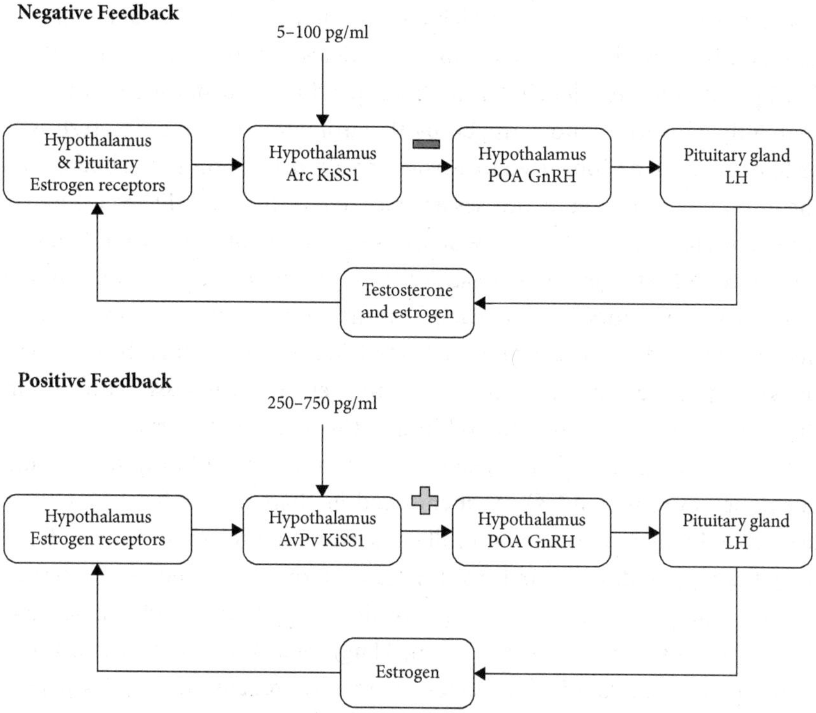

Figure 2.3 Representative homeostatic control of a physiological system. Reproductive physiology is tightly regulated by homeostatic feedback loops in both sexes. Males only experience negative feedback control, whereas females have both negative and positive feedback processes. The function of homeostatic balance is to maintain a relatively constant level of gonadal steroids (e.g., estrogen).

is kept within an optimal range, the set point of which is determined by neuropeptides localized to specific brain regions in the hypothalamus. Here, we will focus on the negative-feedback control of a reproductive steroid hormone, estrogen, in male and female mammals, although the principles described are applicable to the wider amniote lineage. In mammals, negative-feedback functions to maintain estrogen within a relatively narrow concentration range from 10–60 pg/ml in human males and 20–80 pg/ml during the mid-follicular phase of the human female cycles (Figure 2.3). Within the framework of Cannon's homeostatic concepts, estrogen is considered the regulated physiological variable. Estrogen is primarily produced from the ovary, specifically by the outer layers of granulosa cells that surround growing oocytes (Hunzicker-Dunn and Mayo, 2015). In males, the primary source of estrogen is derived indirectly from the testicles (Smith and Walker, 2015). Leydig cells and Sertoli cells in the testes produce testosterone that is subsequently converted into estrogen by the aromatase enzyme. Interestingly, testosterone can be converted to estrogen in almost all cells in the body, even the brain.[1] Any cell that expresses aromatase has the capability to create estrogen by changing the structure of testosterone. In both males and females, if the levels of estrogen increase above the set point, there is greater binding to estrogen receptors (i.e., sensors) that are located on kisspeptin (Kiss1) neurons located in the arcuate nucleus of the hypothalamus. In this case, the kisspeptin neurons function as an error detector. The kisspeptin cells in the arcuate nucleus project to and inhibit the activity of a group of cells in the preoptic area of the hypothalamus that produce gonadotropin-releasing hormone neurons (GnRH). As mentioned in chapter 1 (Table 1.1), GnRH drives luteinizing hormone and follicle-stimulating hormone release from the pituitary gland, which in turn stimulates steroid hormone secretion (i.e., estrogen) and gamete development within the gonads of both males and females. In this example, therefore, GnRH neurons function as the controller of circulating sex steroid hormones. Increased arcuate nucleus kisspeptin activity in response to higher blood estrogen levels result in the inhibition of GnRH release into the anterior pituitary gland. The final component to complete the model for homeostatic balance of estrogen is the effector. Luteinizing hormone cells in the anterior pituitary gland express receptors for GnRH and secrete the hormone into the general circulatory system in response to stimulation from GnRH and effectively serve as the effector. Therefore, the negative feedback of estrogen into the hypothalamus lowers GnRH release from the hypothalamus and results in lower secretion of luteinizing hormone from the anterior pituitary gland and in turn, reduces estrogen production by the gonads in both males and females. In addition to negative feedback, estrogen has multiple distinct effects on the GnRH population (Herbison, 1998), these form a long-term rheostatic, regulated change in activity discussed below.

The major advantage of homeostatic processes is the relatively rapid responses to deviation from the set point. Stability in core body temperature is one classic example of a quick stimulus-response form of homeostasis. Any change from the core body temperature of 37°C or the circulating estrogen concentrations of 10–80 pg/ml are typically returned to the set point within minutes to hours. These examples highlight the fundamental nature of homeostasis to act as an intrinsic physiological mechanism across the animal kingdom for the short-term stability of regulated variables. These innate systems are an evolutionarily ancient processes that can be characterized as the primary stasis.

Physiological stability and a moving set point

Yet, there are several physiological processes in which a set point range for a physiological variable is exceeded without severe pathological consequences or death. Instead, other secondary stasis levels act to maintain physiological balance over longer timescales. Let's return to our examples of regulated variables of core body temperature and circulating estrogen concentrations to highlight the difference between primary homeostasis and secondary rheostasis. Across the day, there is a predictive change in core body temperature in which the set point will reside around 36.2°C during sleep/rest and then peak at approximately 37.6°C during the middle of the active day. This is a controlled temperature span of over 1°C and, in terms of molecular biology, the difference between a highly efficient or poorly driven enzymatic reaction. Another example will occur in little over half of the world's population. During the female lifespan she will experience a predictable change in reproductive physiology in the form of menstrual periods. In mammals, female reproductive periods are generally classified as either estrous or menstrual. The estrous period is the most commonly observed in mammals, whereas menstruation is limited to primates, a few bats, elephant shrews, and the spiny mouse. Both types of female reproductive periods require fine control over multiple circulating hormones such as luteinizing hormone and follicle-stimulating hormone in the anterior pituitary gland as well as estrogen and progesterone from the ovary. Across the menstrual period, circulating estrogen concentrations will span a much larger range (250–750 pg/mL) which is due to a predictive change in the set point established by the hypothalamus. Daily changes in body temperature and female reproductive period are two physiological changes that exhibit robust cyclic changes in set points that cannot be explained by Cannon's concepts of homeostasis. These relatively long-term changes in physiological stability can be differentiated from homeostatic systems at multiple levels and include genes, cellular populations, neuronal circuits/connections, and behavior.

The main physiological change which occurs during the female reproductive period includes a switch between follicular and luteal phases. In humans, the follicular phase includes the time between the first day of the period and ovulation, whereas the luteal phase consists of the time between ovulation and the beginning of the next period. During both follicular and luteal phases, homeostasis in the form of negative feedback is constantly active. Yet, the switch from the follicular phase to the luteal phase involves a marked change in which the regulated variable, estrogen, changes from homeostatic negative to positive feedback. During positive feedback in mammals, a different neuronal population of kisspeptin, located in the anteroventroperiventricular (AvPv) nucleus of the hypothalamus, responds to estrogen levels by stimulating the GnRH neurons in the preoptic area to release more luteinizing hormone from the anterior pituitary gland (Figure 2.3). This process produces a significant increase in estrogen concentration, the purpose of which is to cause ovulation and facilitate uterine receptivity to the implantation of an egg that has been fertilized by sperm, and subsequently impregnation.

A second function of increased estrogen is to stimulate female reproductive behavior by acting in the hypothalamus and the ventromedial nucleus. Professor Donald Pfaff, at the Laboratory of Neurobiology and Behavior at the Rockefeller University, has made several significant contributions to our understanding of how the hypothalamus regulates female reproductive physiology and behavior.[2] Over a research career that has spanned 40 years, some of his more notable discoveries was the identification of estrogen receptors in the brain (Pfaff, 1968) and of the developmental programming and function of GnRH neurons (Schwanzel-Fukuda & Pfaff 1989). In his book *Drive* (1999), there is an exhaustive explanation of how the hypothalamus regulates reproductive behavior in female animals. A succinct overview of the role of sex hormones and brain pathways that regulate negative and positive feedback during the female reproductive cycle is also provided in an excellent review in the scientific journal *Hormones and Behavior* (Pfaff et al., 2008). We will revisit the maintenance of stability in female reproductive physiology in more detail in Chapter 4.

The well-orchestrated changes from genetic-to-behavioral levels that underpin the stability of repeating reproductive periods in females provides a clear example of how the concept of homeostasis cannot be extended beyond relatively short-term, second to minute timescales. So then, how are long-term regulated changes in physiological processes controlled if not through homeostatic mechanisms? In relation to core body temperature, annual changes in the insulative properties of mammalian coats—wool in sheep and fur and hair in other animals such as deer—have been recognized for millennia. Simply, homeostasis in core body temperature primarily acts over the short term to provide a physiological response to external and internal changes to the milieu interieur. But what evidence can be mustered to illustrate that long-term physiological changes, such as those described in the previous chapter, are maintained despite rapid adjustments in homeostatic processes?

Here, the physiological and neuroendocrine control of short-term and long-term cycles in energy metabolism provides a useful framework to highlight the distinction between acute homeostatic control and long-term programs. Siberian hamsters exhibit a remarkable change in body mass across the seasons. In the late spring to early autumn, hamsters are large with an average mass of 42 grams but can range between 35 grams and 55 grams. Then during the winter periods, hamsters will remain in the hibernaculum and will lose roughly 30% of their body mass. Professor Fran Ebling at the University of Nottingham developed an experimental approach to separate short-term homeostatic energy stability and long-term rheostatic stability. A rather simple and ingenious experimental design to test the difference between homeostasis and long-term regulated changes in energy balance is to maintain hamsters in either the summer long-day photoperiod state or a short-day, winter, photoperiod condition. After 8–12 weeks in short days, hamsters will be in a lean "anorexic"-like state compared to the heavier hamsters maintained in the long days (Figure 2.4). So, the hamsters appear to have two stable states and flip between those states because of an internal program that controls circannual timing.

It is important to emphasize that summer and winter conditions change several physiological processes and behaviors involved in energy balance, such as reduced food intake and motivation to eat. Despite the dramatic shift in body mass, the homeostatic control of energy balance is maintained. If hamsters in summer and winter conditions experience a temporary food restriction, such as a 24-hour fast, body mass will decrease in both groups. The short-term homeostatic control of energy balance is activated in the heavier summer and leaner winter hamsters, despite the over-arching long-term seasonal program for larger mass in one state versus the other. Examples such as differences in long-term seasonal variation in body mass

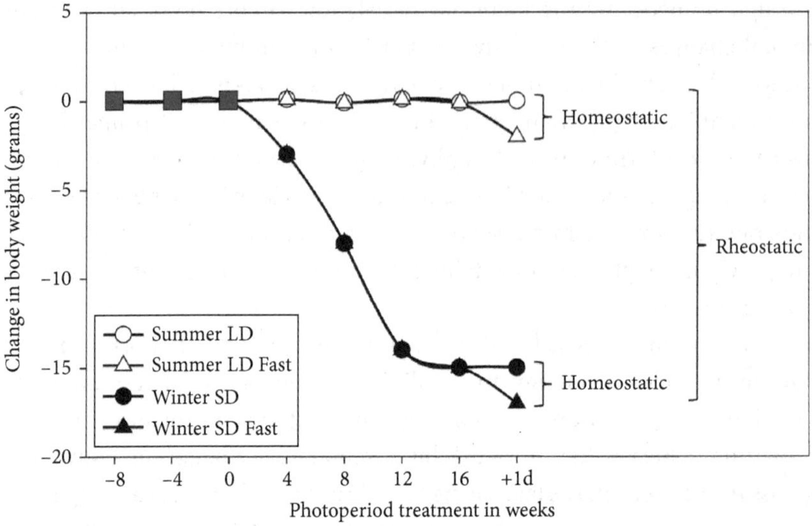

Figure 2.4 A simple example to illustrate the difference between homeostasis and rheostasis.
Siberian hamsters kept in a summer long-day photoperiod of 16 hours light, 8 hours dark will maintain a larger body size (indicated in grey boxes). If hamsters remain in long days, they maintain a relatively constant body mass (indicated in white shapes). However, if hamsters are moved to a winter short-day photoperiod of 8 hours light, 16 hours dark, there is a large decrease in body mass of approximately 30%, from 45 grams down to 30 grams in a matter of 12 weeks. The seasonal photoperiod change in body mass is regulated by a long-term rheostasis mechanism. An easy means to show that homeostasis is maintained in the two seasonal body mass conditions is to fast hamsters for 18 hours (i.e., overnight) (indicated in triangles). Fasted hamsters will lose approximately 2 grams in body mass in both summer and winter conditions compared to animals with constant access to food (indicated in circles). The short-term homeostatic change in body mass is associated with an increased motivation for food intake in fasted animals.

and short-term homeostatic mechanisms clearly indicates that another internal timer governs *steady* adaptive changes over extended time scales. Here the long-term stability in energy balance reflects a programmed change in stability that is best described as *rheostasis*. The distinct genetic, cellular, and hormonal characteristics that differentiate the short-term homeostatic and long-term energy stability will be described in Chapter 5.

Alternative conceptual frameworks for non-homeostatic stability

Several physiological phenomena show predictable, anticipatory, and regulated processes across multiple timescales that cannot be explained by homeostatic processes. One of the earlier attempts to characterize these non-homeostatic physiological changes was developed by Stylianos Nicolaidis (1977) while working in his Laboratoire de Neurobiologie des Regulations at the Centre National de la Recherche Scientifique in Paris. The original term coined was homeorheusis, incorporating *rheursis*, a Greek term describing a system that is flowing or plastic instead of *static*. A problem with using homeorheusis however was that the term had been used by the embryologist Conrad Waddington to describe the programmed development of organisms. Waddington was an embryologist working on the impact of early life environments on the developing chicken. He used homeorheusis to describe how many physiological processes during embryonic development do not respond to cellular, hormonal, or external perturbations. As many of the long-term cycles we aim to describe and explain persist throughout an organism's lifespan, and do not simply reflect the well-defined stages of the developmental trajectory, homeorheusis is not appropriate for describing non-homeostatic physiological change. It is important to note that there have been other attempts to formalize a working model to describe long-term regulated changes in physiology, notably energy balance and body mass. Wayne Kuenzel at the University of Arkansas developed a model to account for the annual change in the autonomic nervous system of migratory birds that could be dissociated from homeostatic systems (Kuenzel, 1999). Kuenzel described the effect of *poikilostasis* (Greek for *various* and *steady states*) and proposed a series of interconnected brain regions called the visceral forebrain system as the primary structures that regulate annual changes in physiology responsible for successful migration. Another model framework, developed by the French-Canadian Michel Cabanac, was the *ponderostat*, which unfortunately did not garner significant attention (Cabanac et al., 1971). One of the key components of Cabanac's model was an emphasis on the role of behavior in contributing to a regulated variable (1987). The disadvantage of these models was primarily the lack of generalization of the concepts beyond energy balance.

Another model of physiological stasis has been described that provided an additional level in the hierarchy: *allostasis*. Toward the later end of the twentieth century, it was becoming increasingly recognized that anticipatory behavioral responses could act by "*achieving stability through change*" (Sterling and Eyer, 1988). The concept of allostasis recognized that environmental challenges posed by unexpected or life-threatening situations would induce a behavioral response that, in part, served to maintain physiological stability. An excellent example to demonstrate allostatic responses is the stress response. Several have been books written on the physiological and behavioral basis of allostasis and the consequences of chronic damage to an individual's ability to cope with change. A few excellent resources are *Allostasis, homeostasis and the costs of physiological adaptation* (Schulkin, 2015), *Why zebras don't get ulcers* (Sapolsky, 2004), and *The end of stress as we know it* (McEwen and Lasley, 2002).

It has long been recognized that an individual or animal's response to a particular environmental stimulus, such as the presence of a predator or final exam, leads to heightened levels of anxiety or emotional states. Many of the behavioral responses elicited by an individual are based on prior experience with the same stimuli, or related stimuli, and are driven by brain regions outside those involved in homeostasis (i.e., hypothalamus). Instead, a completely separate physiological system acts to provide "top-down" or behavior-dependent stability that occurs in response to an environmental-induced change in internal states. Here, the concept of allostasis explains how higher-order cortical and limbic structures can augment physiological responses and seek to achieve stability through change (Sterling and Eyer, 1988). Allostasis has also been proposed to explain the maintenance of physiological stability across seasonal life history stages (McEwen and Wingfield, 2003). While some of these arguments have merit, recent scientific advances, and the ability to isolate specific cell populations in the brain that function to maintain stability during periods of change, independent of behavioral state, support the conjecture that distinct physiological systems are involved in the control of these regulated long-term changes.

A model that incorporates both homeostasis and allostasis into a single framework, the reactive scope, was developed by Michael Romero and colleagues at Tufts University in Massachusetts (Romero et al., 2009) and includes both *predictive homeostasis* and *reactive homeostasis*. In this model, predictive homeostasis incorporates all physiological processes that are cyclical (e.g., daily and seasonal). Reactive homeostasis is defined by the range of a physiological variable required to respond to unpredictable environmental stimuli. Any response outside the upper and lower limits of the physiological range of the reactive scope are considered *homeostatic overload* or *homeostatic failure*. The reactive scope model has been useful to develop a single framework for physiological stability, especially for responses such as the stress response. The advantages and limitations of allostasis will be discussed in further detail in Chapter 8.

Formal properties of regulated stability: defining Rheostasis

It was a contemporary of Michel Cabanac who developed formal terminology that is appropriate for the study of regulated control of physiological processes. Nicolas Mrosovsky argued for the use of the root *rheo* which is the Greek word for flow or stream and with stasis defines changes in a steady state. The key distinction between rheostasis and homeostasis is the ability to incorporate the definition of "regulated changing levels" to differentiate from "homeo" which indicates "same" and implies a rigid constancy. Other terms used to encompass cyclical or regulated changes in set points have used *predictive homeostasis* as mentioned above. However, this terminology is imprecise and results in a major confounding issue when examining neural pathways and the genetic basis of a physiological model. The primary limitation of predictive homeostasis is that the term implies that the characteristics of homeostatic systems can simultaneously engage in short-term responses (i.e., stimulus-response, negative feedback) and coordinate long-term responses. One aim of this book is to highlight that homeostatic systems do not have the capacity to independently "predict" or anticipate changes in the environment. The evidence presented is that an entirely separate neuronal and genetic system has evolved to coordinate regulated changes in physiological processes. This book will continue to use the term rheostasis as described by Mrosovksy and attempts to development a singular framework to account for long-term regulated changes in physiological systems that occur independent of homeostatic mechanisms.

Mrosovsky developed a rheostatic model based on the homeostasis characteristics described by Cannon. For Mrosovsky's model, instead of a central set point, rheostatic systems had at least two distinct sensory/detector steps that could maintain the upper and lower level of a regulated variable (Figure 2.5). The current model of rheostasis was proposed over 30 years ago and needs an update. This is largely in part due to the major advances

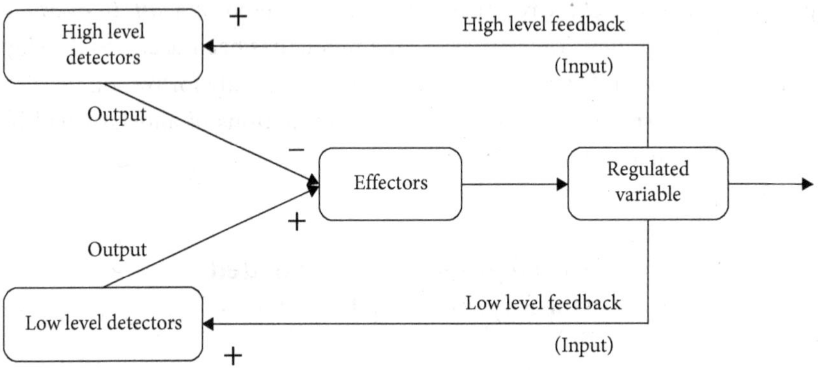

Figure 2.5 The model of rheostasis proposed by Nicholas Mrosovsky. Rheostasis includes many components similar to homeostatic control of a physiological variable. There are 3 fundamental differences: (1) the absence of a set point, (2) the combination of sensors and detectors, and (3) the switch from an isolated controller to a "control-system."
Figure adapted from Mrosovsky, 1990.

in the genetic and neuronal control of homeostatic systems and complex neuron connection tracing methods in murine animals that have facilitated the identification of a highly interconnected nervous system. However, the Mrosovsky model provides a useful framework to start to untangle the neuroendocrine control of long-term physiological stability. Yet, a couple pitfalls are evident. First the absence of a central controller that acts to govern the switch from low to high and vice versa is required. Second, as outlined in Chapter 1, endogenous timers such as circadian clocks are integral components that need to be included. These two limitations highlight that additional neuro-anatomical and/or genetic concepts need to be incorporated into Mrosovsky's rheostasis model. Rheostasis works well to describe several physiological processes, and here the example of seasonal cycles in Siberian hamster energy balance provides a starting point to understand the basic structure of the model. Seasonal cycles in the hamster body as a regulated variable has long-term changes that constantly feed back into the rheostatic mechanism. The summer long photoperiods maintain a higher body mass and the high-level detectors in the hypothalamus inhibit effector systems (e.g., adipose tissue) to maintain heavier hamsters. In short-day photoperiods there is a decrease in body mass that signals to the low-level detectors. The output signal from a low-level detector in turn stimulates effectors leading to an increase in body mass. This switch from high-level detectors to low-level detectors and back could be envisioned to govern the circannual timer. Using well-defined examples of rheostasis, the subsequent chapters will seek to expand on the model proposed here. Before we start to delineate an updated model of rheostasis that uses scientific advances over the past few decades, we need to outline some key features of long-term regulated changes in physiology.

The formal experimental investigation of rheostasis had significantly stalled over the past 30 years but is now well positioned to make substantial advances. Until the early 1990s, basic and medical research had predominantly focused on the physiology and cellular properties associated with homeostasis and pathological conditions. Since the 1990s the ability to generate genetic "knock-out" mice, such that a single gene was deleted from the genome, was a major step forward to understanding the genetic basis of rheostatic control of physiological systems, behavior, cognition, and disease. For rheostasis, there are two types of long-term processes: those that cycle with a regular periodicity, such as daily, tidal, lunar, and seasonal physiology; and physiological responses that are regulated over several days to months, and even annually, but do not oscillate. The former, long-term changes in the set point for a regulated physiological variable that include predictive cycles are classified as *programmed rheostasis*, whereas physiological processes that adhere to a well-defined length of time, but occur only once, such as acute infections or pregnancy are referred to as *reactive rheostasis*. In Mrosovsky's book, the examples used to illustrate programmed rheostasis included body mass cycles during hibernation, migration and female reproductive cycles, temperature across the female reproductive cycle, seasonal reproductive cycles, antler growth, acid-base status during hibernation, and circadian and sleep cycles. Mrosovsky's examples to illustrate reactive rheostasis included fever, infection-induced anorexia, injury, hypothermia, dieting, and prolonged feeding on high fat diets. For the purposes of the perspective presented in this book, pregnancy and lactation, which Mrosovsky originally postulated as programmed, are considered as reactive rheostatic processes.

Hierarchical organization of physiological stability

Table 2.1 provides an overview of the similarities and differences for homeostasis, rheostasis, and allostasis. The key distinction is that rheostasis

Table 2.1 Summary of the hierarchical control of physiological stability

Form of Stasis	Biological Organization	Evolutionary Organization	Time scale	Anatomical basis	General function	Damage or disruption
Homeostasis (primary)	Innate	Conserved	Seconds to minutes	Central nervous system (i.e., hypothalamus) & Autonomic nervous system	Essential for life	Life threatening
Rheostasis (secondary)	Innate	Species-specific	Hours to days	Central nervous system (i.e., hypothalamus)	Maintenance of physiological change	Chronic pathology
Allostasis (tertiary)	Experiential	Individual-dependent	Seconds to days	Central nervous system (i.e., limbic and cortical)	Maintenance of physiological change	Chronic pathology

provides a framework in which changes in regulated physiological variables are controlled over extended time to optimize an individual's fitness and conditions for a healthy life. To reiterate, homeostatic mechanisms are intrinsic and an evolutionary conserved process that is observed across chordates. The primary distinction of homeostatic processes is the short-term course of action with physiological stability restored from seconds to minutes. The rapid response is in part enforced due to the autonomic nervous system engagement for sensory detection and effector responses. Any loss of function associated with homeostasis would lead to life-threatening situations and likely death. Rheostatic mechanisms can be dissociated from homeostatic processes at multiple levels. First, rheostasis has innate properties, such as a genetic underpinning, but the physiological response may not be conserved across all animals and instead can show species-specific responses. The rheostatic control of a regulated variable occurs slowly and requires hours to days and even months to ensure physiological stability at an organismal level. Unlike homeostasis, disruption to rheostatic responses can be accommodated, yet there are long-term deficits that are associated with chronic pathologies. Finally, allostasis is a tertiary stasis in the hierarchical organization of physiological stability. Here, an individual's prior experience provides guided input on subsequent exposure with an environmental factor that causes an internal shift in a regulated variable. Unlike homeostasis and rheostasis, there is strong neuronal input from limbic structures and cortical regions that act to fine tune hypothalamic control of physiological responses. Any damage to these brain structures and the loss of allostatic control does not necessarily lead to a life-threatening physiological state but can lead to long-term or chronic pathologies that significantly impair an individual's physiological stability.

Summary

In this chapter, the focus has been to define rheostasis and place the long-term stability in physiological processes into a conceptual framework. It is clear that a hierarchical organization of physiological balance is maintained across the animal kingdom and can be divided into primary, secondary, and tertiary levels. Programmed rheostatic changes, including cyclic variation during the day, or reactive rheostatic associated with pregnancy occur independent of cognitive or affective states. Chapters 3–5 will describe examples of programmed rheostasis whereas Chapters 6–7 detail reactive rheostasis. For each example, the genetic, cellular, neuronal, and neuronal network for rheostatic processes will be outlined and evidence presented for how their regulation differs from Canon's foundational level of homeostasis. The overall aim for the next series of chapters is to develop a new conceptual framework for rheostatic processes that build upon the homeostatic model described by Walter B Cannon.

Notes

1. The idea that estrogen is a female hormone is a complete myth. Estrogen is found systemically in the blood and produced locally in many cells in males as well as females. Estrogen is critical for creating a "male-like" brain. To learn more about the role of estrogen to establish sex differences or the creation of a male brain, readers should review McCarthy, 2008.
2. On my way to the Annual Society for Behavioral Neuroendocrinology conference in Mexico, I had the pleasure of sitting beside Professor Donald Pfaff on a flight from Houston to Querétaro. He was an incredibly welcoming individual to discuss research projects and as an early career researcher at the time, I gained helpful insight into the thought process of a brilliant scientist.

3

An endogenous clock for programmed rheostasis

Oft a little morning rain foretells a pleasant day.

Charlotte Brontë (1816–1855) "Life" (1846)

In Chapter 2, we saw how our understanding of homeostasis centers around quick responses that are generally used to hold a physiological variable in its optimum working range. Cannon recognized that homeostasis could not explain all natural physiological variation and stated: "Unfortunately we know very little about the stimuli which cause these slow adjustments of the organism that protect against an unfavorable or harmful environment." The example Cannon focused on was long-term changes in hair growth in hibernating animals, which they use to adapt to the cold harsh winter periods but then lose in the spring. One could equally state that physiological systems function to anticipate a favorable environment to increase an organism's fitness. The significance of long-term regulated changes in physiological stability which occur over daily timescales was not appreciated until later in the twentieth century.

Evidence for programmed daily internal timers

We now appreciate that daily cycles in physiological stability is an evolutionary ancient internal timer and consists of conserved genetic and molecular pathways. Many plants exhibit day-night changes in flowering in which their petals close during dusk and engage in "sleep" movements, a process known as nyctinasty. Charles Darwin was particularly interested in nyctinasty and sleep in plants. In collaboration with his son Francis, he conducted several experiments on nyctinasty and concluded that light intensity was a

On Rheostasis. Tyler John Stevenson, Oxford University Press. © Oxford University Press 2024.
DOI: 10.1093/oso/9780197665572.003.0003

powerful environmental cue regulating flower petal movements and growth (Darwin and Darwin, 1898). Their work also showed that the absence of a regular period of light resulted in less leaf damage after plants were treated with radiation. However, over a century before, the French astronomer Jean-Jacques d'Ortous de Mairan (1678–1771) had conducted a rather ingenious study in which he kept a heliotrope plant (assumed to be *Mimosa pudica*) inside a box that was deprived of any light information. De Mairan monitored the leaf movements of the plant for several days and discovered that an intrinsic rhythm in movements was observed with an approximately 23-hour period (de Mairan, 1729). This means that what Darwin saw and described as a daily change linked to light exposure happened even in the absence of light and therefore that there must be an internal timer that can match or predict changes in the timing of sunlight. Furthermore, there must be some interaction between this internal timer and the environmental cues that coordinate the cycle in leaf movements with the daily waxing and waning of the sun. Findings such as these highlighted that internal processes were engaged *prior* to a future environmental condition and thereby provided an anticipatory change in plant physiology.

Evidence for these endogenously driven circadian cycles is widespread and is seen in plant, insects, and vertebrate species. Yet in the not-so-distant past, some researchers argued against a genetic basis for circadian timing and instead supported a role of experience or a "learned" component. In the early 1900s, Richard Semon proposed that circadian timing was due to light-dark cycles experienced by an individual (or parent) that induced persistent physiological rhythms in plants and other animals (Semon, 1904), such that the 24-hour cycle (e.g., 12hr light, 12hr dark) in leaf movements would be maintained in plants after they were moved to constant light/darkness environments. Here, the light-dark cycle induced an experience-dependent physiological response that continued in the absence of light information. He maintained that an oscillatory motion, akin to the waves produced in water after a rock was dropped in a puddle, had been established in the plants. The logic here was that the internal rhythms produced in constant conditions should reflect the previous environment (i.e., 12hr light, 12hr dark). If this theory was correct, one would expect a 12-hour rhythm in plants that experience 6hr light, 6hr dark. However, plants that are exposed to a 6hr light, 6hr dark light schedule and then moved to constant darkness (or light) will exhibit a near 24hr rhythm in leaf movement! These findings that the rhythm in the plant (or animal) persists with an approximate 24-hour period when a cyclic daily environmental cue is manipulated such that it repeats faster (or slower) than a day supports the hypothesis that an internal program drives the 24-hour rhythm, which is completely separate from experience or learning.

Semon continued to argue against endogenous rhythms and proposed that rhythms could also be inherited based on the experience of the organism's parent(s). Again, however, his hypothesis is not supported by data. If the inheritance perspective was correct, then hundreds of generations living in a constant environment would be predicted to induce a loss in circadian rhythmicity. This question was examined using a common circadian invertebrate model—the fruit fly (*Drosophila melanogaster*). Even when bred in constant light for over 600 generations, flies will have offspring that still display clear and robust circadian cycles in locomotor activity, similar to flies bred under control conditions of oscillating light dark (Sheeba et al., 1999). Similar results have also been reported in mice, although in that case the breeding program only extended to 7 generations (Aschoff, 1955). Altogether, these types of data establish that endogenous circadian rhythmicity is genetically based, completely autonomous, and is neither based on experience nor inherited. For an excellent overview of the history of the field of circadian biology, the reader is encouraged to seek the chapter by Serge Dann in *The circadian clock* (2010).

Daily cycles in regulated physiological variables

Daily cycles can be visualized using a typical sine-wave form (Figure 3.1). These sine waves will have "peaks" and "troughs" that occur at a single point in the day. But across the day, there are specific points, called the *phase* of the daily rhythm, for example, the point at which an animal wakes or falls asleep. Regulated physiological variables maintain 24-hour cycles, with a peak, a trough, and at each point in the waveform can be classified by a specific phase (Gamble et al., 2014). For example, blood hormone concentrations are tightly regulated by negative-feedback homeostatic processes, yet there are predictable peak levels, such as cortisol peaks early in the morning phase. Other hormones that peak in the morning include the two gonadotropes, luteinizing hormone and follicle stimulating hormone. Thyroxine and insulin, hormones associated with metabolism, show peak levels during an afternoon phase and troughs during the very

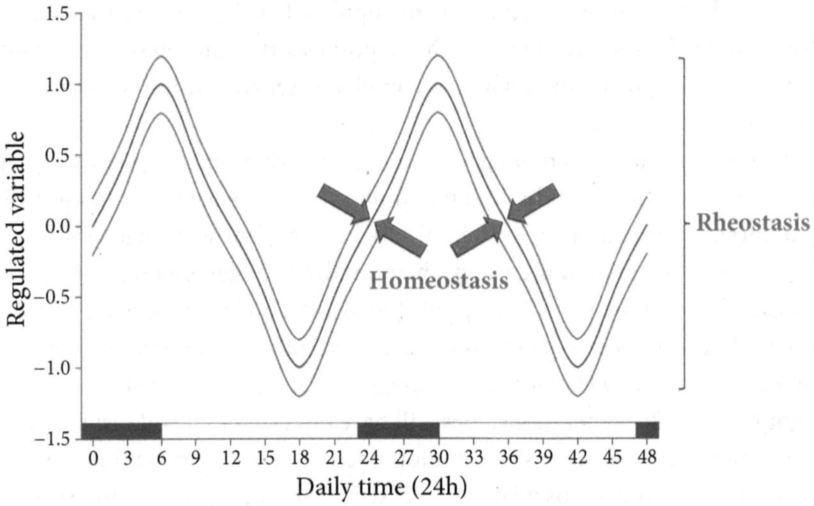

Figure 3.1 Daily programmed rheostasis and homeostasis.
A schematic representation of daily rhythms in a regulated variable (e.g., cortisol) is indicated by the black line. The rhythm is plotted against the mean daily value of the regulated variable. Homeostatic responses to deviation from a set point occur rapidly and are generally consistent across the day. Upper and lower thresholds for the activation of homeostatic systems are indicated in red. The red arrows indicate increased or decreased changes in regulated variables to maintain internal stability. Daily programmed rheostasis is the regulated change in homeostatic set points and includes endogenous circadian clocks located in sensors, controllers, and effectors.

early morning phase. A few, but critically important hormones, show peak concentrations during the dark phase; these include growth hormone, melatonin, and prolactin (Halberg et al., 1983). Even the molecules that were the focus of Walter B Cannon's research display regulated changes in blood concentrations across the daily cycle. Salt balance requires the co-ordinated control of both sodium and chloride concentrations within the body, and both show peak blood plasma concentration in the late evening phase, whereas urine levels peak in the early afternoon phase. In Chapter 2, we discussed Jurgen Aschoff (1983) and the identification that core body temperature in humans is also under the control of an endogenous circadian clock, with increasing temperatures during the day and a subsequent decrease of about 2°C during the dark phase. Data such as these indicated that even classically homeostatic regulated physiological variables can be dependent on the time of day. Moreover, the daily variation across diverse physiological systems is not synchronous, which indicates individually programmed changes in the set points, controllers of physiological variables, and/or levels of tissue specific changes established by the effectors. Therefore, each regulated variable is controlled by short-term homeostatic responses and programmed rheostatic responses that provide anticipatory changes to coordinate optimal physiological states across an organism to match the time of day.

Dissociation of homeostasis and programmed rheostasis

But how can we dissociate homeostatic and rheostatic responses to establish independence in mechanisms of establishing physiological stability? If homeostatic systems were the only means to maintain stability, then any deviation from a set point should return to a specified value regardless of the time of the day. Yet, daily rhythms in regulated variables are pervasive and show large differences in peak and trough values (Figure 3.1). In humans, the peak in plasma cortisol is 25 ug/dl and occurs shortly before an individual awakes; the trough occurs late in the day at 5 ug/dl. The increase in morning cortisol is essential to activate key reticular activating centers in the brain and is critical for the control of arousal from a sleep state. Throughout the day, the hypothalamo-pituitary-adrenal axis (HPA) maintains a relatively constant level of cortisol (i.e., glucocorticoid) production from the adrenal cortex. Negative feedback systems in the paraventricular hypothalamus (i.e., corticotropic-releasing hormone) regulate the blood concentrations of corticosteroids and is actively engaged at all times of day (i.e., all phases) (Nader et al., 2010). The predictable change in cortisol set points is a strong indication that all homeostatic systems: sensors, error-detectors, controllers, and/or effectors change to accommodate homeostatic responses. Examples such as plasma corticosteroid hormone concentration highlight the difference between homeostatic and rheostatic systems. The acute, short-term deviation from the daily phase in cortisol value is consistent across the day, yet the regulated set point is under strong *circadian* control. Here, we will consider studies that illustrate the distinction between homeostatic and rheostatic responses, how programmed circadian rheostasis functions with homeostatic systems, and how the nervous system codes for programmed changes in regulated values.

One could argue that the daily programmed change in physiological variables is a simple response to the environmental conditions, similar to Richard Semon's criticisms of the genetic basis of the circadian clock. To illustrate an endogenous change in a regulated set point, the astute researcher needs to "challenge" the physiological system being examined. Let us consider energy balance. One way to examine the homeostatic regulation of blood glucose is to conduct a simple *glucose tolerance test*, which consists of measuring blood glucose before a meal and then collecting samples for several hours after to determine how an individual removes consumed glucose from their circulatory system. If a homeostatic set point is maintained across the day, then post-prandial (after the meal) plasma glucose concentration should return to the exact same concentration regardless of the time of day. If a rheostatic set point is imposed to regulate daily changes in blood glucose, then one would predict that the shape of the glucose response curve and/or the return to "baseline" plasma glucose concentrations would vary with time of day. Work conducted in Frank Sheer's laboratory in the Division of Sleep Medicine at Harvard Medical School examined plasma glucose profiles after the exact same meals were given to participants to be consumed in the morning (i.e., 0800h) or evening (i.e., 2000h) to understand how circadian timing effects energy balance. After both meals, there is an increase in blood glucose from approximately 80 mg/dl, yet the morning peak in glucose was only 160 mg/dL compared to 192 mg/dl after the evening meal, an average difference of over 30 mg/dL! Additionally, the time to return to pre-meal glucose concentrations was, on average, 30 minutes quicker in the morning compared to the evening (Morris et al., 2015). These data indicate that the homeostatic response is quickly activated after a meal, as the glucose concentration is returned to the pre-meal level. But a rheostatic program that is dependent on the time of day functionally alters the systems that regulate physiological variables (i.e., plasma glucose). Blood plasma concentrations of cholesterol and triglycerides maintain daily rhythms with peak levels in the early evening. Glucose on the other hand has peak blood plasma concentrations early in the morning, prior to the light phase. The glucose tolerance test is a reliable assay of the stability of blood glucose concentrations and is consistently observed across the animal kingdom. For example, a similar response occurs in mice. To test daily variation in glucose balance, mice were fed 6 meals a day; each meal lasted 10 minutes and were spaced 4 hours apart. As predicted from a programmed rheostatic model, the change in plasma insulin was time-of-day dependent (Kalsbeek and Strubbe, 1998).

Meals that were ingested during the dark phase elicited a larger increase and longer duration of plasma insulin while mid- and late-light phase meals induced a smaller increase in plasma insulin. The positive challenge to energy stability using food intake provides a useful example to illustrate that those daily challenges to a physiological variable (e.g., plasma glucose/insulin) are under both homeostatic and rheostatic control. We will return to long-term programmed rheostasis in energy balance again in Chapter 5.

Genetic basis of internal timers that control daily programmed rheostasis

Now that we have established that homeostatic and rheostatic mechanisms operate to regulate physiological systems across each 24-hour period, we need to consider the mechanisms through which daily programmed rheostasis is integrated with homeostatic regulation. The key concept here is that a genetically determined circadian clock is embedded within homeostatic systems and the coordinated control of daily clocks across sensors, set points, error detectors, controllers, and effectors provides the programmed rheostatic control of regulated variables. What do we know about the circadian clock at a genetic, cellular, and behavioral level? Several other beautifully articulated books and scientific reviews have been published on the molecular mechanisms that control daily rhythms, and the reader is encouraged to consult those resources for an in-depth understanding (see Chapter 1). Let us first consider the genetic basis and transcriptional-translational feedback loops that establish cell- and tissue-autonomous 24-hour time and how the circadian clock functions to maintain long-term internal stability.

The fruit fly (*Drosophila melanogaster*) has been a valuable insect model to study the genetic and molecular basis of daily rhythms. The significance and great benefit of the fruit fly is the capability to change the structure of the genome in specific genes and cause mutations that stop the function of a gene or genes of interest. Manipulations in the fly genome have helped to identify the genetic basis of daily cycles in physiology and behavior. Ronald Konopka, working as a graduate student with Seymour Benzer, was the first to illustrate that mutations in an unknown gene completely abolished normal daily locomotor behavior cycles of flies (Konopka and Benzer, 1971). Konopka used ethyl methanesulfonate, a mutagenic compound that randomly substitutes nucleic acids, to induce genetic mutations in the fruit fly genome. He then conducted behavioral tests in mutant flies to identify those that maintained daily light responses in eclosion rhythms and, using a clever classical genetic analysis that included backcrossing mutant flies, showed that a single genetic locus caused impairments in daily behavioral rhythms. The duration of the light insensitive part of the cycle was 5 hours shorter in mutant flies than in wild-type or untreated flies. As the length of eclosion rhythms shortened the active part of the daily cycle in mutants, Konopka named the gene period (*Per*). His work was fundamental in establishing that genes controlled daily rhythms in behavior and laid the foundation for the next generation of scientists. During the 1980s, two groups of molecular biologists, including Jeffrey Hall and Michael Rosbash at Brandeis University and Michael Young at Rockefeller University, independently determined that the gene *Per* was critical for the function of endogenous daily timing of fly locomotor behavior (Reddy et al., 1984; Baylies et al., 1986). Their research showed that the concentration of the protein product of *Per* increased in flies in anticipation of the dark phase, and the levels would cycle over a near 24-hour period. Mutations in the *Per* gene result in shorter (i.e., 19hr) or longer daily cycles (i.e., 28hr) or completely arrhythmic behavior in flies. Their pioneering research resulted in them being awarded the Nobel Prize in Physiology or Medicine in 2017. Unfortunately, as Ronald Konopka had died in 2015, he could not share in the award. The *Per* gene has now been characterized across the animal kingdom and is present in insects, amphibians, birds, and mammals, including humans.

Since their pivotal work, Hall, Rosbash, and Young as well as several other groups have characterized a series of circadian clock genes. For example, the discovery of the gene clock (*Clock*) in mice by the research group led by Joseph Takahashi at the University of Texas Southwestern Medical Center was instrumental for understanding that multiple circadian clock genes were present in the genome and that these genes form a complex, interacting system that is critical for maintaining daily cycles in animal health and wellness (Vitaterna et al., 1994).[1] The genes involved in regulating circadian physiology and behavior have been shown to form a complex transcription-translational feedback loop system that provides most cells with a 24-hour clock (Figure 3.2). During the light phase two circadian clock genes, *Clock* and brain and muscle ARNT-Like 1 (*Bmal1*), are translated into protein and move from the nucleus to the cytoplasm. The protein products then combine or dimerize and are relocated back into the nucleus where they initiate the expression of multiple other genetic targets. The ability of CLOCK-BMAL1

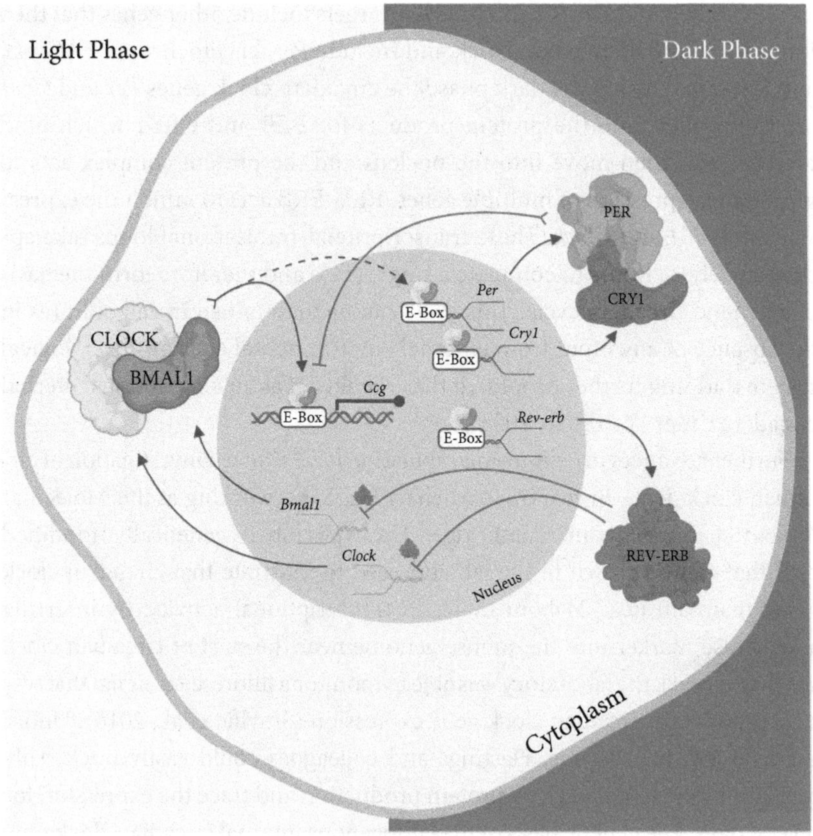

Figure 3.2 Circadian transcriptional-translational clock.
A collection of genes that include period (*Per*), cryptochrome (*Cry*), brain and muscle ARNT-like 1 (*Bmal1*), circadian locomotor output cycles kaput (*Clock*), and Rev-erb form a complex interconnected pathway that maintains an endogenous oscillation that approximates a 24-hour daily clock. The function of the circadian clock is to regulate the daily timing of circadian clock genes (*Ccg*). One of the primary drivers of the circadian clock is the binding of Bmal1-Clock proteins to E-Box motifs in the promotor regions of gene to stimulate transcript expression.

action to specifically regulate daily rhythms in gene transcription from across the entire genome is achieved by a small genomic motif referred to as an E-BOX. Genes that have an E-BOX located in the promoter region, an area slightly in front of a gene sequence, can bind with CLOCK-BMAL1, after which the CLOCK-BMAL1 complex initiates the expression of circadian clock-controlled genes (*Ccg*). Three key targets include other genes that then form the internal circadian clock and include *Per*, cryptochrome 1 (*Cry1*), and Rev-erb. During the dark phase the circadian clock genes *Per* and *Cry1* are transcribed into the protein products for PER and CRY1 which bind together and then move into the nucleus and the protein complex acts to inhibit the expression of multiple genes. REV-ERB acts to inhibit the expression of *Bmal1* and *Clock*. These transcriptional-translational loops take approximately 24 hours to complete a single cycle and therefore form the basis of a genetic circadian cycle. This cycle can be maintained in cell cultures in the absence of any input from external environmental cues or physiological systems, adding further proof that these genes act as an endogenous internal circadian timer (Yoo et al., 2004).

Further advances in genome manipulation have allowed investigation of circadian clock genes in real time. Michael Hastings, working at the Molecular Biology Laboratory in Cambridge, UK, developed genetically modified cells that can be grown in the lab and used to illustrate that circadian clock genes maintain near 24-hour cycles in transcriptional activity. By inserting a luciferase marker into the mouse genome near the start of circadian clock genes (e.g., *Per*), the laboratory was able to monitor a fluorescent signal that was a by-product of circadian clock gene expression (Smyllie et al., 2016). Under a fluorescent microscope, Hastings and colleagues could easily track daily oscillations in circadian clock protein production and trace the expression for several days. The genetic basis of the endogenous internal circadian clock now understood, it was possible to investigate how its dysfunction affects rheostasis and homeostasis. For example, it has been shown that mice that harbor a mutation in the *Bmal1* or *Clock* gene have very different responses in the regulation of energy balance compared to control (normal mice). Mice with mutations in the circadian clock genes will likely have higher daily glucose concentrations and elevated daily waveforms in glucose and triglycerides (Rudic et al., 2004; Turek et al., 2005). However, when mice with mutations in circadian clock genes were challenged with an injection of insulin there was an immediate activation of the homeostatic mechanism that worked to return plasma glucose concentrations to pre-challenged levels. Note that wild-type control mice returned to a stable state in energy balance more quickly in comparison to *Bmal1* and *Clock* mutants. These findings highlight that the circadian clock system maintains endogenous daily rhythms in glucose concentrations, and any disruption to the integrity of time keeping directly impacts programmed rheostasis, although homeostatic responses remain intact.

Interestingly, not all cells that express 24-hour cycles contain the circadian clock transcriptional-translational feedback loop. For example, mature mammalian red blood cells are *enucleated*, meaning they do not contain a nucleus, so they cannot use the same genetic timing system to maintain 24-hour rhythmicity. Instead, an alternative internal timing program is present. John O'Neill and Akhilesh Reddy working at Cambridge University discovered a series of conversed antioxidant proteins, known as peroxiredoxins, within red blood cells, which show approximately 24-hour cycles in chemical reactions in which oxygen is removed from molecules. These chemical reactions are maintained in the absence of environmental and internal signals (O'Neill and Reddy, 2011). Rachel Edgar and colleagues (2012) subsequently established that peroxiredoxin proteins are conserved across bacteria, archaea, and eukaryote lineages and may provide a common cellular circadian timekeeping mechanism.

It is now well established that all living animals have a cluster of genes that serve to provide cells with an endogenous 24-hour clock. Several research laboratories have discovered that the circadian clock gene network functions in a cell- and tissue-specific manner to drive daily changes in circadian clock–controlled genes to optimize or regulate cell function in a time-of-day dependent manner. This circadian clock is responsible for the daily rhythms in multiple physiological variables and, therefore, is a major driver of programmed rheostatic changes to maintain daily internal stability. We now need to consider and understand how the circadian clock controls daily programmed rheostasis, though we need to first understand how the circadian clock is fine-tuned by external signals.

Neural pathways and hormone signals involved in daily programmed rheostasis

Across the plant and animal kingdoms there is considerable variation in which external environmental signals can influence the functioning of the circadian clock. Across many systems, natural as well as artificial light can modify, or *entrain*, the endogenous cycling of the circadian clock to match the 24-hour period of the environment. Even though most cells contain the circadian clock system, light information does not directly regulate the timing of gene expression. There are two ways in which light can impact the genetic basis of circadian cycling in cells located throughout the body: either by light-induced regulation of a hormone signal or by the activation of neuronal pathways linked to the integration of light information that subsequently impacts the transcriptional-translational circadian clock. In most cases, both hormonal and neuronal systems are involved in the light-dependent regulation of the circadian clock and subsequently programmed changes in regulated physiological variables.

In mammals, the eyes contain the sensory tissue required for light detection and critical for establishing neuronal communication of light. The light receptor melanopsin, also referred to as a photoreceptor, is localized to light-sensitive retinal ganglion cells (pRGC) in the retina of the eye. Initially identified in light sensitive skin cells collected from the African clawed frog (Provencio et al., 1998), these exact same photoreceptors have been subsequently characterized in most mammals and shown to be critical for light-dependent entrainment of daily rhythms (Freedman et al., 1999). The presence of light activates melanopsin cells and the information is transmitted by the retino-hypothalamic nerve tract to the master circadian nuclei, the suprachiasmatic nucleus (SCN) within the hypothalamus (Figure 3.3). The light stimulation in the SCN then acts to entrain the endogenous circadian clock genes in neurons to maintain the functional integrity of the SCN and internal 24-hour daily cycles. The SCN neurons directly connect to several other hypothalamic nuclei; one of the major afferent projections is to the paraventricular nucleus (PVN). The PVN contains many neurons that synthesize corticotropic-releasing hormone (CRH) and so the SCN-PVN connections provides the CRH neurons with light/dark stimulation. As the CRH neurons are the head of the HPA, this connection is used to regulate programmed daily rhythms in cortisol concentrations, outlined above. The widespread connections between the SCN and other hypothalamic nuclei allow the SCN to coordinate and influence daily cycles in neuroendocrine function across many physiological systems.

A fundamental disadvantage of neuronal circuits is that they require a connection: in most cases many connections are required between the central nervous system and any effector tissue/cell. The SCN pathways are supplemented by an indirect circuit in which the SCN clock regulates the production of melatonin from the pineal gland. In all animals, the hormone melatonin is produced by the pineal gland in the brain. It provides the main physiological variable that codes for night length and is transmitted to the entire body via the circulatory system. As a result, melatonin could be referred to colloquially as the "hormone of darkness." The duration of melatonin

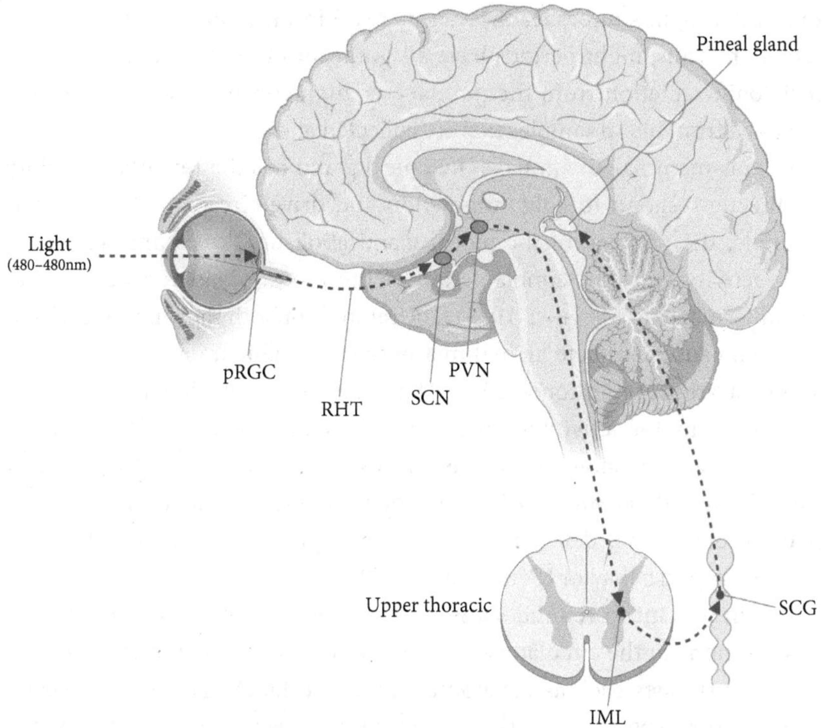

Figure 3.3 Neuronal pathway for light-dependent daily rhythms in mammals. Light wavelengths between 460 nm and 480 nm activate melanopsin in photosensitive retinal ganglion cells (pRGC) in the eye. Light activation is passed down the retinohypothalamic tract (RHT) into the suprachiasmatic nucleus (SCN) in the hypothalamus. The SCN projects to the paraventricular nucleus (PVN), and the PVN cells project into the spinal cord down to the intermediolateral cell column (IML). There, sympathetic postganglion cells originating in the superior cervical ganglion cell (SCG) project to the pineal gland to inhibit the synthesis of melatonin.

secretion provides a reliable physiological signal that most animals use to determine seasonal day length, or more suitably night length. The SCN-PVN link is critical for melatonin production, as it signals daylight via a convoluted pathway that includes cells in the intermediolateral cell column (IML) of the spinal cord and subsequent sympathetic post-ganglion innervation to the pineal gland. The pineal gland therefore is the critical brain tissue for the organismal-wide communication of light information. Light stimulation has a dramatic inhibitory effect on melatonin production and secretion: the removal of light sources leads to an increase in melatonin production and secretion. Thus, night periods have a significant increase in the amount of melatonin secretion from the pituitary, as high concentrations are readily obtained from blood samples that are collected at night.

In non-mammalian animals, the pineal gland has the remarkable ability to directly respond to light stimulation and drive circadian rhythms. For example, the removal of the pineal gland abolishes "free-running" circadian activity and the reimplantation of the pineal gland restores activity (Zimmerman and Menaker, 1979). The removal of both the pineal gland and eyes can cause complete arrhythmia in pigeons (Ebihara et al., 1984). The administration of melatonin using a daily regimen is sufficient to restore circadian locomotor activity in birds (Gwinner and Benzinger, 1978). Similar results have been shown in lizards and fish (Underwood, 1989). It is now well established that multiple brain regions in non-mammalian animals have photoreceptors that detect the presence of light to drive daily changes in physiology and behavior (Pérez et al., 2019).

It is important to remind some readers that the synthesis and presence of a hormone in the circulatory system does not equate to a functional outcome. Hormones such as melatonin can act on target tissues and cells by binding to receptors and activating an intracellular pathway (e.g., cAMP) that induces an indirect change in gene transcription leading to RNA expression. Alternatively, hormones may cross into a cell and bind to intracellular receptors which together bind to the genome to induce gene transcription leading to RNA expression. Melatonin has multiple receptors, and the activation of each receptor results in slightly different functional outcomes. In humans, the two most studied melatonin receptors are MT1 and MT2, both with several different isoforms, such as MT1a, MT1b, and MT1c. The US National Institutes of Health (NIH) maintains the National Center for Biotechnology Information, a library of electronic information that is vital for biomedical and basic research and contains the vast wealth of scientific knowledge for research.[2] A simple search for the human MT1a receptor (https://www.ncbi.nlm.nih.gov/gene/4543) results in a treasure trove of basic and clinical information relating to its location and function. Analyses of different tissues has revealed that the MT1a is expressed in tissues including the brain, kidney, colon, and pancreas. MT1b (https://www.ncbi.nlm.nih.gov/gene/4544) is also expressed in the brain but has higher expression in other tissues such as the placenta, lymph nodes, and fat. So, there is a significant level of overlap between the expression patterns for the receptor isoforms.

But the wide tissue distribution means that daily changes in light and dark periods can be conveyed to the entire body, providing a mechanism through which to enact/maintain programmed rheostasis.

Dissociating homeostatic and programmed rheostatic systems

We have now established that an endogenous daily clock resides in animal and plant cells and external cues such as light can influence the endogenous clock to maintain day-to-day cycles. Early examples using energy balance illustrated that homeostasis maintains constant internal stability, yet many regulated physiological variables exhibit anticipatory, programmed changes across the day. But what happens when the circadian clock genes or melatonin signaling are lost or defective? One would predict that damage to either circadian clock genes or melatonin pathways would result in disruption in programmed rheostasis (i.e., the circadian clock). But homeostatic mechanisms should not be impaired.

Mice models are extremely valuable in studying the possible effects of disrupted rheostasis, as we can induce specific mutations in the circadian clock genes of mice and subsequently study the animals to ascertain how physiological systems are affected. Here circadian clock gene mutations provide an opportunity to separate rheostatic and homeostatic regulation of the internal milieu. In mice, blood glucose has a circadian rhythm that maintains high levels during the early morning, which declines across the day (Rudic et al., 2004). Mice that have a mutation in the circadian clock gene *Bmal1* maintain homeostatic balance, although with disrupted responses. For example, following an injection of glucose or insulin directly into the body cavity (i.e., intraperitoneal) there is a physiological deviation away from the glucose set point and a subsequent activation in a homeostatic response that includes storage and breakdown of circulating glucose. In wild-type mice, an injection of insulin results in an immediate decrease (i.e., within 30–60 minutes) in blood glucose, the size of which does not change across the day (roughly 0.25 mg/dL decrease). These data indicate that under normal genetic conditions, the homeostatic balance is constant despite the large daily changes in set points. In mice that have either a *Bmal1* or *Clock* mutation in the gene sequence, have a larger decrease and longer recovery time in blood glucose (Rudic et al., 2004). Targeted tissue ablation of *Bmal1* in the liver of mice resulted in an increased blood glucose concentration in *Bmal1* compared to wild-type mice after both glucose and insulin injections (Lamia et al., 2008). Glucose and insulin tolerance responses in *Bmal1* knockout mice also require additional time (approx. 50 minutes) to return to pre-injection levels. Here the sensory detector, error correction, controller, and effectors for homeostatic balance are entirely functional and induce a physiological response. The caveat is that the regulated physiological response to the set point is lacking, which highlights that the rheostatic input from the circadian clock (i.e., SCN) provides an additional layer of control for internal stability.

Similarly, MT1 and MT2 gene knockout mice maintain homeostatic changes in energy balance albeit with slightly augmented responses. Compared to wild-type mice, glucose injections lead to a greater change in blood glucose (approximately 30mg/dL) and an hour longer to recover in MT1 knockout mice. In response to an insulin injection, MT1 knockout mice have a delayed homeostatic response in blood glucose concentrations (Owino et al., 2018). There was little to no difference between MT2 and wild-type controls in blood glucose concentration in response to intraperitoneal injections of glucose (glucose intolerance test) or insulin (insulin tolerance test) (Contreras-Alcantara et al., 2010). These data indicate that MT1 may provide the primary melatonin-dependent organism-wide signaling for daily programmed rheostasis.

Both *Clock*, MT1, and MT2 mutations also have a significant effect on long-term indicators of energy balance: body mass. The absence of these genes in mouse genomes results in an increase in body mass for *Bmal1*, *Clock* (Rudic et al., 2004), and *MT1* (Owino et al., 2018). These findings support the notion that prolonged disruption of circadian rhythms in energy balance have detrimental effects for the maintenance of body mass, despite an intact homeostatic system. One limitation to this interpretation is that genetic mutations, particularly in mice, are limited in their ability to extrapolate findings to natural gene-environment interactions with relevance to humans. One way to study the role of programmed daily rheostasis is to induce (or experience) transient yet long-term disruptions to daily rhythms in animals with the full complement of circadian clock and melatonin receptor genes.

The capability to disrupt rheostasis while maintaining homeostasis is essential to illustrate that the two physiological processes can be differentiated. Here, circadian disruption can be used to demonstrate that long-term challenges to programmed rheostasis can lead to pathological conditions, yet homeostasis can remain unaltered. Several good examples that are derived from human research on circadian disruption include shift work, environmentally induced jetlag, and socially induced jetlag. A study conducted at the Centre for Clinical Investigation at Brigham and Women's Hospital examined the postprandial glucose and insulin responses to an isocaloric diet given at 0800h and 2000h in outwardly healthy volunteers who worked chronic shift patterns. The key experimental manipulation was the inclusion of a simulated workday in which the circadian system was aligned, compared to an intervention consisting of simulated night work that included a circadian misalignment resulting from 12-hour inversion of behavioral and environmental cycles. Both circadian phase and circadian misalignment groups exhibited a clear homeostatic response with an acute elevation in blood glucose and insulin concentrations after their meal at both 0800h and 2000h. As expected, the circadian phase impacted the shape of the homeostatic response, with a prolonged elevation in glucose after the evening meal. This altered return to baseline blood concentrations emphasizes that programmed rheostatic input from the SCN was significantly impaired. In shift workers that experience the misalignment of their workday relative to the environment (i.e., those working the night shift), there was a slight increase in blood glucose concentrations. Postprandial blood insulin levels were only affected after the evening meal in the misaligned treatment group (i.e., shift workers) compared to the shift workers in the circadian alignment group. These data highlight that glucose intolerance is maintained regardless of circadian (mis) alignment, but a disruption to shift workers' schedules reduced glucose tolerance (Morris et al., 2016).

Conceptual framework for daily programmed rheostasis and homeostasis

How does programmed circadian-driven rheostasis integrate with the homeostasis model outlined in Chapter 2? As described in the previous chapter, homeostasis is maintained by several components including sensors, set points, error detectors, controllers, and effectors that serve to regulate a specified physiological variable. There are two means by which programmed rheostatic control of physiological systems interact with the homeostatic components. First, each homeostatic component has a circadian clock that imposes time-of-day dependent responses to deviations from the set point

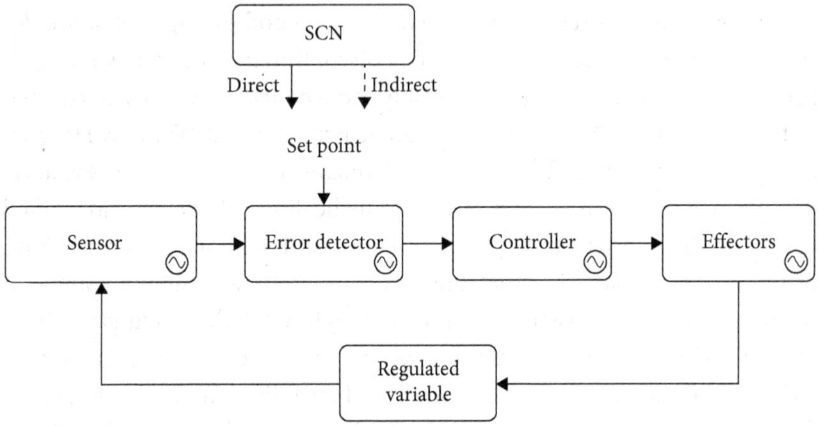

Figure 3.4 Schematic model of programmed rheostasis with circadian clocks. Physiological systems show daily changes in homeostatic set point values. The daily maintenance of internal stability is regulated by circadian clocks that reside in sensors, error detectors, controllers, and effectors. The SCN provides direct and indirect input to regulate daily cycles in set points.

(Figure 3.4). Second, for any physiological set point, the master pacemaker nucleus, the SCN, conveys two signals to the integration center which provides information about environmental (daily) time. The direct pathway includes connections via the nervous system, and the indirect pathway is coded using the hormone melatonin. The direct and indirect inputs provide the time-of-day information that is used to regulate daily cycles in the set points of physiological variables. The SCN provides the basic component for programmed rheostasis to govern the shifting daily set point. In addition to the influence of SCN signaling, the circadian clock genes ensure daily timing in peripheral structures involved in the sensing, error detection, and controlling homeostatic responses. A disruption to any rheostatic component (i.e., SCN or circadian clock genes) will result in a long-term deficit in programmed rheostasis that is manifested in pathological conditions, leading to chronic illnesses.

Summary

In this chapter, we have seen examples of daily changes in mammalian energy balance using glucose and insulin tolerance tests, which were used to highlight the difference between homeostatic and rheostatic systems. We have also seen that there is a distinct genetic network intrinsic to most cells in the body, which acts independently of the homeostatic response. Instead, the chapter argues, the circadian clock system drives daily variation in regulated physiological variables. An intact SCN, a discrete brain region of the hypothalamus, is essential for daily programmed rheostasis; damage to this region results in a complete loss of anticipatory changes in physiology. Finally, the hormone melatonin functions to synchronize physiological variables across tissues and endocrine systems. The circadian clock, SCN, and actions of melatonin provide distinct genetic, neuronal, and hormonal substrates that maintain rheostasis and act entirely independently of the genes, neuroanatomical centers, and hormones involved in homeostatic error detection, controller, and effector responses. This rheostatic system provides a higher order system in which the SCN provides an additional regulatory layer that is critical to allow anticipatory changes to the internal state and maximize physiological efficiency. The predictability of rheostatic systems optimizes any given physiological response to a daily change in the environment and thereby increases an animals' fitness. In Chapter 4 we will focus on female reproductive cycles and build on the evidence for a distinct rheostatic system that has genetic, cellular, neuroanatomical, and behavioral components that can be separated from the homeostatic control of reproductive physiology.

Notes

1. On the February 20, 2017, I had the opportunity to visit the Taj Mahal with Joseph Takahashi and Ken Wright. At the time I had just started my laboratory in the School of Biological Sciences at the University of Aberdeen. During one of the tour breaks I asked Joe which of his scientific discoveries was his most favorite (as there are many), to which he replied, "the characterization of the clock gene."
2. There is a world of gratitude for the US taxpayer. The NIH is predominantly supported through tax revenues and these funds are used to maintain a public database of information that includes scientific publications and knowledge of life sciences. The NCBI database is essential for animal (including human) health and welfare for the entire world. The global population is a major beneficiary of the information and service provided by the US citizenry.

4

Orchestration of female reproductive cycles

> Nothing in the world is single; All things by a law divine.
> In one spirit meet and mingle. Why not I with thine?
>
> Percy Bysshe Shelley (1792–1822),
> "Love's Philosophy" (1819)

All organisms on earth share one fundamental characteristic: reproduction. Across the three domains of archaea, bacteria, and eukaryote, organisms must replicate either through asexual or sexual reproduction to maintain the survival of the species (Smith JM, 1978). The alternative is death and extinction. Asexual reproduction involves only one parent, and all copies (or offspring) are genetically identical to each other and can be considered clones. Asexual reproduction includes binary fusion (i.e., bacteria and archaea), vegetative propagation (e.g., plants derived from stems, leaves, and roots), the production of spores (e.g., fungi), fragmentation (e.g., yeast), and parthenogenesis (e.g., daughter clones of adult female aphids). Conversely, sexual reproduction involves the joining of two gamete cells, a process referred to as fertilization. In this case, there are two parents that are not genetically similar and produce offspring that have a unique genetic make-up compared to the parents. Examples of sexual reproduction include syngamy, or the fusion of two gamete cells into a single zygote, the most common type of reproduction in multicellular populations and widespread in birds and mammals. Other single-celled organisms such as bacteria, protozoans, and fungi can reproduce sexually via a process known as conjugation in which a cytoplasmic "bridge" is formed between two cells, forming a temporary fusion for the exchange of genetic material.

The advantages of sexual reproduction are the increased genetic variability in the population, increased ability to adapt to new environments, and decreased likelihood a disease will eliminate the population. The

On Rheostasis. Tyler John Stevenson, Oxford University Press. © Oxford University Press 2024.
DOI: 10.1093/oso/9780197665572.003.0004

disadvantages are the time and energy required (i.e., find a mate) and the possible inability to propagate, in the event an individual is isolated on a deserted island. Advantages of asexual reproduction include rapid increases in the population, as only one individual is required, there is more time to reproduce as less effort is spent finding a mate, and self-replication is incredibly energy efficient. There are disadvantages though, such as the lack of genetic variation in the population. An organism is only suited for a limited, if only one, environment, and a higher susceptibility to disease. In a very limited number of species, both systems are used, sexual reproduction being preferred but under certain conditions asexual reproduction being possible.

Both asexual and sexual reproduction are tightly regulated processes that requires optimal timing. In Chapter 3 we outlined how programmed circadian rheostasis is present across physiological systems to time daily functions. The timing of reproduction, both asexual and sexual, is also influenced by the circadian clock genes. In cyanobacteria, the circadian clock is a critical component of reproduction that constantly monitors cell division to increase rates during the dark phase (Martins et al., 2018). A functional circadian clock is also involved in the daily timing of births in a range of mammals such as mice (Miller et al., 2004), and rats (Reppert et al., 1987). Work by Peter Martin and colleagues (2018) which looked at the timing of over 5 million human births in England over a 10-year period identified that under natural conditions most births occur between midnight and 0600. However, when we look at timing of regulated variables of reproductive physiology, we see that it also consists of cycles organized over longer time scales, such as female ovarian cycles over days to weeks and annual/seasonal cycles of fertility in some animals.

Similar to the brain-centered role of programmed circadian rheostasis described in Chapter 3, it had long been known that discrete brain regions were involved in the control of female reproductive physiology. But the specific genes that control the pituitary gland and hormone synthesis by the gonads took longer to identify. In the 1940s it was becoming clear that the hypothalamus contained "releasing factors" that controlled the secretion of pituitary hormones. George Harris was one of the early pioneers in neuroendocrine systems and proposed that these releasing factors were centrally involved in the homeostatic regulation of reproduction (Harris, 1948). It was not until two decades later that Schally and Guillemin used thousands of hypothalami dissected from pigs and sheep, and obtained from abattoirs, to isolate "luteinizing hormone release factor" (Amoss et al.,

1971), a brain-derived hormone that regulates luteinizing hormone release from the anterior pituitary gland. The hormone was subsequently called gonadotropin-releasing hormone (GnRH) as it stimulates the release of both gonadotropins: luteinizing hormone and follicle-stimulating hormones (Schally et al., 1971). The discovery of neuropeptides that govern the release of hormones from the anterior pituitary gland resolved the identity of the primary "controllers" for the homeostatic regulation of physiology and led to Andrew Schally and Roger Guillemin being awarded the Nobel Prize in Physiology or Medicine in 1977 (Wade, 1981).

As with the circadian regulation of physiological systems, rheostatic control can be separated from the homeostatic balance to maintain physiological stability of gonadal hormones (e.g., estrogen) within an optimal range in a minute-by-minute basis. Timing female reproductive cycles requires a range of hormone changes over the course of the cycle and involves multiple brain regions. The neural control of female reproduction and behavior has been studied in detail for decades. Using rodent models, Donald Pfaff and a plethora of students, research associates, and other colleagues at the University of Rockefeller uncovered the brain pathways that regulate female reproductive physiology and behavior. Pfaff, a graduate of the Massachusetts Institute of Technology, has held a professorship and worked as head of the Laboratory of Neurobiology and Behavior since 1978. His early research characterized the ability of sex steroids, testosterone, and estrogen to bind to specific brain regions, such as the limbic systems and hypothalamus, in both male and female rats (Pfaff, 1968). An excellent overview of the neuroendocrine regulation of female reproductive behavior is outlined in a comprehensive review in the *Hormone and Behavior* journal (Pfaff et al., 2008). Over the past 20 years, advances in gene technologies have allowed the characterization of genes involved in the control of female reproductive stages and revealed a complex system devoted to the long-term, rheostatic regulation of reproductive physiology.

In this chapter we will briefly cover the homeostatic regulation of gonadal steroids before considering the role of programmed rheostasis in timing the female reproductive cycle. The primary focus will include mammalian reproductive cycles and draw from estrous and menstrual cycles. Cycles of egg-laying and incubation in birds will be used to highlight the conserved neural and cellular control of reproductive cycles. The chapter will build on the fact that reproductive timing involves programmed circadian rheostasis and that timing estrous/menstrual cycles also include another, higher-order timer called central pattern generators.

Homeostatic regulation of female reproductive hormones

In Chapter 2 we discussed the role of negative and positive feedback for the regulation of estrogen in human females, and the homeostatic regulation of blood concentrations that range from 20 pg/ml to 80 pg/ml during the mid-follicular phase in humans. Figure 4.1 provides the homeostatic model for the regulation of estrogen. In brief, estrogen receptors act as the sensor to detect circulating levels of estrogen. Any deviation from the 20–80 pg/ml set point range during the mid-follicular phase results in a correction by estrogen receptor expressing kisspeptin neurons in the arcuate nucleus. Once a deviation from the set point is detected by this kisspeptin neuronal population an error signal is transmitted from the kisspeptin cells to the evolutionary conserved "controller" of reproduction: GnRH neurons in the preoptic area. GnRH neurons are the ultimate controller of reproductive physiology, as these cells project into the median eminence to stimulate the gonadotropin cells in the anterior pituitary gland to release luteinizing hormone (LH). LH is released in proportion to GnRH activity, moves through

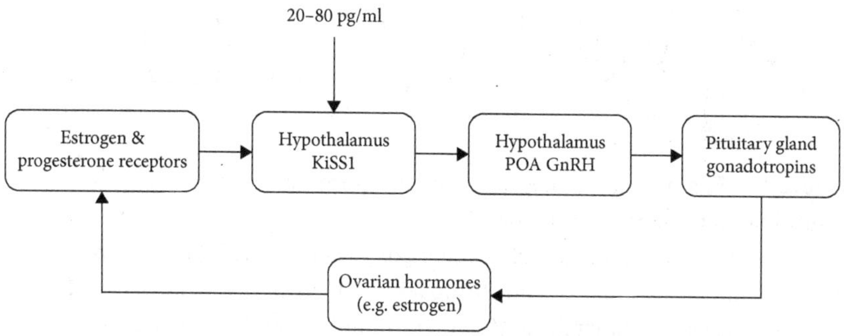

Figure 4.1 Homeostatic regulation of female reproductive physiology. During the mid-follicular phase of the human menstrual cycle the levels of blood estrogen are maintained at 20–80 pg/ml. Kisspeptin (*Kiss1*) neurons in the arcuate nucleus monitor blood estrogen concentrations and signal to the GnRH neurons in the preoptic area (POA). GnRH neurons are the master neuropeptides that control reproductive physiology in all sexually reproducing species. These neurons control the release of two gonadotropins from the pituitary gland. One of these, luteinizing hormone, has a direct effect on the ovary to produce estrogen. An increase in blood estrogen results in the inhibition of luteinizing hormone release from the pituitary gland, which results in less estrogen synthesis by the ovary.

the circulation system to the gonads (i.e., ovaries), and binds to LH receptors in theca cells in the ovary to subsequently trigger the production of estrogen in adjacent granulosa cells.

Studies that investigate the homeostatic regulation of reproductive physiology predominantly use a scientific approach that includes conditional tests of necessity and sufficiency. Tests of necessity seek to establish whether the change in regulated variables always follow removal of a particular effector (e.g., gonads). For example, LH levels in the circulatory system in rodents is typically maintained around 2 ng/ml (Figure 4.2A). Upon the removal of the gonads (i.e., necessity) there is a significant increase in plasma LH concentrations, a nearly tripling in level to 6 ng/ml. If the product of the gonads (e.g., testosterone) is replaced using a subcutaneous implant, there is a substantial decrease in LH levels and a return in homeostatic balance of

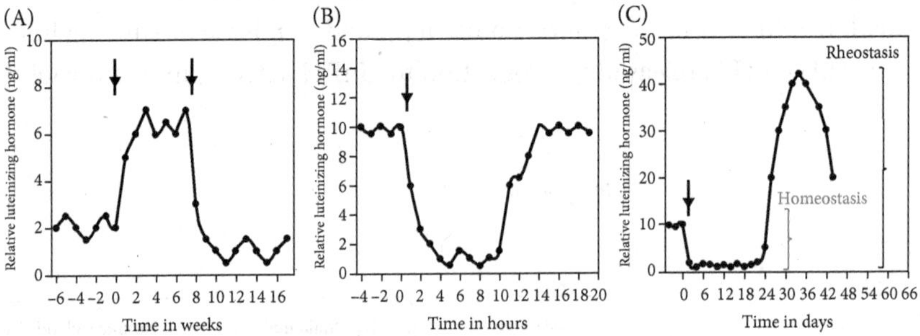

Figure 4.2 Examples of homeostasis and programmed rheostasis for the control of reproductive physiology.
Gonadal hormones tightly regulate luteinizing hormone (LH) release from the pituitary gland using homeostatic feedback mechanisms. (A) In males, LH is maintained at approximately 2ng/ml in the blood. Castration indicated by the first downward arrow results in significantly higher LH blood concentrations that can be maintained for several weeks. Implants of testosterone indicated by the second downward arrow lowers LH concentrations. (B) In females, a single injection of estrogen can rapidly reduce LH concentrations (<1hr). After several hours the inhibitory feedback is gradually dissipated due to break down of estrogen hormone, and the LH concentrations return to regulated levels. (C) At a specific time during the reproductive cycle injections of estrogen initially inhibit LH concentrations, but then a programmed change in hormone sensitivity results in positive feedback leading to an LH surge. This programmed switch from inhibitory to stimulatory effects of estrogen is a clear example of long-term regulated changes associated with rheostasis.

approximately 2 ng/ml. The administration of testosterone indicates the sufficiency of the hormone to maintain homeostatic control of LH. Studies that use gonadectomy, both castration and ovariectomy, coupled with hormone replacement require prolonged periods of time due to the number of weeks required to eliminate endogenous levels of the regulated variable (i.e., gonadal steroid). In females, the concentration of LH circulating in the blood is significantly greater compared to males, and in rodents, LH is regulated at around 10 ng/ml. A quicker approach to test the sufficiency of effectors to change the concentration of a regulated variable is a single injection of estrogen in an intact animal followed by monitoring of the blood concentrations of the regulated levels of LH for several hours (Figure 4.2B). In this case, a single bolus injection in female rodents can trigger an immediate decline in LH levels that persists for up to 10 hours. Once the excess estrogen hormone injected into the animal is degraded, the LH concentrations return to homeostatic balance. These are just two examples that highlight the effect of exogenous gonadal hormones for the homeostatic regulation of blood LH concentrations. But a single injection of estrogen in females can have a very different effect on LH concentrations that is entirely dependent on the stage of the cycle (Figure 4.2C). Estrogen injections in females can initially activate a negative feedback mechanism that reduces LH concentrations. However, if exogenous estrogen levels remain in circulation after 22–26 days, LH concentrations significantly change and increase by up to 40 times! The long-term regulated change in LH concentrations in this instance is a direct manifestation of a programmed rheostatic change in reproductive physiology. The increase in LH concentrations is referred to as the LH-surge and functions to stimulate ovulation and proactive sexual behavior in females. The LH surge is a fundamental feature in both menstrual and estrous cycles in mammals and serves to coordinate the timing of reproductive behavior to increase the probability of successful fertilization of mature gametes (i.e., eggs). Now we will turn to the constellation of hormones that change across estrous and menstrual cycle and use a comparative approach to highlight the evolutionary conserved program.

Beyond homeostasis: naturally occurring cycles in female reproductive physiology

In nature, the usual stages of female reproductive physiology are (1) puberty, (2) follicular development, (3) pregnancy, (4) lactation, with stages 2–4 repeated until death or until the animal has used all available gametes. The observation of female reproductive cycles, either estrous or menstrual, is the consequence of females not getting pregnant. The absence of fertilization by male gametes results in the termination of follicular development and the start of a new reproductive cycle.

The timing of reproduction in female mammals is characterized by programmed cycles of LH surge activity that induce changes in ovarian activity. The two most commonly studied reproductive cycles are estrous and menstrual cycles. In both instances the cycles show a remarkable degree of regularity that can only come about because of an exquisitely programmed physiological system. Most women of reproductive age (e.g., 12–52 years) experience a near monthly cycle, called a menstrual cycle, from the Latin *menstrualis*, "of or having monthly courses," as their period approximates 28 days. Other mammalian species exhibit an estrous cycle, from the Greek and translated as "frenzy," and relating to the period of time during which the female is receptive to mating. The association with a frenzy may have to do with solicitous behavior by the female and competitive behaviors among fertile males.

The temporal control of female reproductive cycles is diverse across mammals. The full cycle consists of both a luteal and follicle stage. The luteal stage occurs after the ovary releases an egg (i.e., ovulation) and ends prior to the shedding of the uterine lining (i.e., menses). The luteal stage can be considered as the first day of the menstrual cycle and continues to approximately 14 days after ovulation (Figure 4.3A). The follicular stage, approximately 14 days before ovulation, is the time between the start of menses and the beginning of ovulation. In other primates, the menstrual cycle can last longer, such as in orangutans (29 days), gorillas (30 days), and chimpanzees (37 days). Other mammals, including rodents, have shorter estrous cycles that last 4–5 days (Figure 4.3B). The estrous cycle in the female rat is one of the best characterized and understood reproductive cycles due to the robust change in hormone profile, availability of large sample sizes (compared to mice), and intense scientific study and associated publication literature amassed for over a century.

Even though the length of the menstrual and estrous cycles varies considerably in mammals there is a remarkable degree of consistency in the programmed rheostatic change in blood concentrations of several hormones

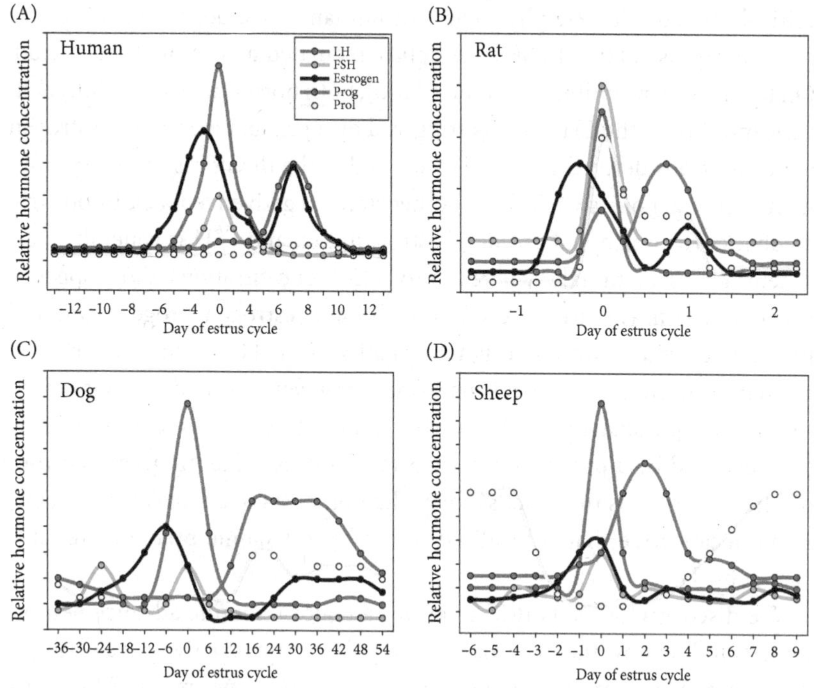

Figure 4.3 Programmed rhythms in hormone concentrations across the mammalian female reproductive cycle.

The female reproductive cycle is conserved across mammals and includes a constellation of hormones produced either in the pituitary gland or ovary. Luteinizing hormone (LH) and follicle-stimulating hormone (FSH) are released by the pituitary gland and stimulate the production of estrogen and progesterone (Prog) by the ovary. Prolactin is also involved in the female reproductive cycle and is released from the pituitary gland. The female cycle is characterized by a distinct phase in which LH and FSH peak. There is considerable diversity in the length in the cycle across mammals.

In humans the length is roughly 28 days (A), 4–5 days in most rodents (B), 90 days in dogs (C), and 16 days in sheep (D). Note that sheep are seasonal breeders and diestrous (i.e., infertile) for most of the year. Hormone data are oriented so that the peak LH/FSH is day 0 to compare hormone rhythms across four mammals selected to illustrate conserved programmed rheostasis in female reproductive physiology.

critical for reproductive physiology. In humans, rats, dogs, and sheep, the gonadotropins (LH and FSH) are tightly regulated across the time period. On the days surrounding ovulation (day 0), both hormones show a "surge" in concentrations. The LH surge is triggered by a gradual increase in estrogen concentrations derived from the ovary and is the direct result of a spontaneous, endogenous switch from homeostatic negative feedback to positive feedback. The hormonal profile of estrogen, LH, and FSH is generally consistent across mammals, whereas prolactin concentrations show species-specific patterns (Figure 4.3A–D). In addition to estrogen, progesterone and prolactin provide convergent signaling pathways that fine-tune the timing of the reproductive cycle. The mechanisms that govern the endogenous switch to positive responsivity of LH to circulating estrogen concentrations is itself governed by a programmed rheostatic process. The programmed rheostatic controller is the exact same as the homeostatic controller and is our small decapeptide that is localized to the hypothalamic preoptic area nucleus: GnRH.

The discovery of the GnRH gene was a major scientific advancement in understanding how the hypothalamus regulates reproductive physiology, as it is essential for reproduction in all sexually reproducing species. Mutations in the gene render an animal infertile and incapable of producing offspring (Gibson et al., 1997). GnRH cells originate in the olfactory bulb during early embryonic development and migrate into the hypothalamus during prenatal development (Schwanzel-Fukuda and Pfaff, 1989). In the adolescent and adult human brain, there are approximately 1000 GnRH neurons dispersed throughout the anterior hypothalamus with higher numbers present in the preoptic area.

GnRH stimulates cells within the pituitary gland that secrete the gonadotropins. Two main characteristics underlie this functionality: a constant pulsatile rhythm and, as described above, control of the "LH surge." In male and female animals, GnRH pulsatile release is important for homeostatic regulation of reproductive physiology, while a GnRH surge is essential for triggering the LH surge. The GnRH surge ultimately results in the release of a follicle from the ovary and increasing estrogen and progesterone concentrations that drive female sexual desire. Disruption of normal pulsatile GnRH secretion is associated with several reproductive disorders including polycystic ovary syndrome (PCOS), hypogonadotropic hypogonadism, and hypothalamic amenorrhea (Christou et al., 2017). However, GnRH cells do not directly respond to ovarian hormones, estrogen in particular. This lack of a direct link between circulating gonadal steroids (i.e., estrogen) and GnRH activity indicated that another neuronal system must provide an intermediate step between the detection of estrogen blood concentrations and the *controller* GnRH. It was not until the end of the twentieth century that the conundrum of error detection and set points for long-term hormonal regulation of GnRH cells for the control of reproductive physiology in mammals began to be resolved.

Mammalian female reproductive cycles involve a genetic KISS

In the late 1990s researchers identified in a family from Saudi Arabia a genetic mutation that resulted in idiopathic hypogonadotrophic hypogonadism, a condition in which there is a deficiency in GnRH release and a subsequent lack of pubertal development. Using complex genome-wide scans of family members and comparison of their genetic sequence with unaffected matched controls, Stephanie Seminara and colleagues discovered mutations in a single G-protein coupled receptor, called GPR54, as the cause of their infertility (Seminara et al., 2003). The question being asked then shifted to: Which ligand (i.e., hormone) binds to the GPR54 receptor to drive pubertal development? Several years before in the town of Hersey, Pennsylvania, a gene had been discovered that inhibited metastasis in a cancer cell line (Lee et al., 1996). Breaking away from the common nomenclature practice of the time the group named the newly discovered gene after a type of chocolate candy made by local company Hershey's, and so the gene was called kisspeptin. In order to establish that kisspeptin was the endogenous ligand for GPR54, Sophie Messager and colleagues conducted a series of elegant studies to demonstrate that kisspeptin injections resulted in elevated LH and FSH, but there was no effect in GPR54 knockout mice (Messager et al., 2005). The identification of kisspeptin and GPR54 receptor provided a major advancement in our understanding of how the set point for GnRH neurons controls pubertal development, and also rheostatic regulation of female reproductive cycles. Yet, in the absence of hormonal feedback, neuropeptides can maintain cyclic activity driven by intrinsic electrochemical properties.

Central pattern generators: cell autonomous timer in neuroendocrine cells

Neuropeptides cells, such as GnRH, form complex neuronal networks that function as a unit. In 1961, a revolutionary concept was proposed that altered our understanding of how neuronal networks could function. Using the flying behavior of the locust (*Schistocerca gregaria*), Donald Wilson discovered that the electrical activity of neurons involved in the motor control of wing movements could maintain rhythmic activity independent of any sensory input (Wilson, 1961). The continued activity of cells that controlled wing movement, without sensory input, showed that some cells can function to control physiological and behavioral processes, endogenously, over prolonged periods. This study was fundamental in establishing the concept of *central pattern generators*: a discrete group of neurons can produce predictable rhythmic outputs over a defined period. We now know that central pattern generators can be found across the animal kingdom, residing in the central nervous system, and rhythmic processes do continue in the absence of sensory feedback or other rhythmic neuronal input (Katz, 2016).

Examples of the central pattern generator systems include feeding, breathing, walking, swimming, and flying. Professor Eve Marder is the Victor and Gwendolyn Beinfield Professor of Neuroscience at Brandeis University and a member of the US National Academy of Sciences. Her work has been seminal in elucidating how microcircuits control the homeostatic regulation of intrinsic membrane properties and how neuronal circuits in the brain remain stable while allowing plasticity to occur during development and learning. An excellent overview of the evolution of central pattern generators is available in *Current Biology*, in particular the paper by Marder and Bucher (2001). Our understanding of central pattern generators is primarily derived from electrophysiology analyses of complex motor behaviors, yet neuropeptides such as GnRH share many properties with central pattern generators because they are released in a pulsatile manner. Rhythmic outputs (e.g., LH surge) of neuronal networks (e.g., GnRH surge), are a direct readout of the central pattern generators and provide a powerful system to make inferences about the underlying control systems.

Neuropeptides in the hypothalamus have a common characteristic that is unique among neurons: these cell types are continually active and maintain neurotransmitter release in a pulsatile manner. Neuropeptides that govern essentially all physiological processes such as oxytocin, vasopressin, corticotropin-releasing hormone, and GnRH show pulsatile activity (Leng and Brown, 1997; Lightman and Conway-Campbell, 2010). As pulsatile activity is a common feature to neuropeptides, what is the functional significance on downstream cell populations in the anterior pituitary or peripheral tissues? Early studies on the neuroendocrine control of regulated physiological variables demonstrated that the constant presence or secretion of a neuropeptide could not induce pulsatile secretion of targeted pituitary cells (e.g., LH). For example, monkeys in whom the GnRH neuronal population had been destroyed and had an intermittent infusion programmed to mimic the natural GnRH pulsatile activity of one injection per hour were able to re-establish the pulsatile activity of gonadotropin secretion (Belchetz et al., 1978). However, if monkeys were administered continuous intravenous GnRH infusions, pulsatile LH secretion was not re-established. Findings such as these indicate that continuous neuropeptide release results in *desensitization* and a decrease in the response of the target cells. One factor that contributes to the down regulated responsiveness of target cells is the decline in the number of receptors responsive to the neuropeptide (Hinkle and Tashjian, 1975).

GnRH neurons are intriguing as they show two forms of intrinsic activity: typical pulsatile release and a surge pattern of release (Herbison, 2016). Precise neuronal recordings from select cells in mouse models have shown that GnRH neurons release the peptide in a pulsatile manner, which increased secretion of molecules every 20 minutes (Czieselsky et al., 2016). But what regulates or drives this intermittent secretory activity? Recent work conducted by Alan Herbison's group—spanning the Centre for Neuroendocrinology at the University of Otago and the Department of Physiology, Development and Neuroscience at the University of Cambridge—has revealed that the neuropeptide kisspeptin plays a role in the control of GnRH pulsatile activity. Using complex neuron stimulation methods called optogenetics, Herbison's group discovered that kisspeptin cells in the arcuate nucleus show increased pulsatile activity that is synchronized across cells and almost perfectly correlated with pulsatile LH secretion from the pituitary gland. The activation of kisspeptin cells in the arcuate nucleus could also induce pulses of LH release (Clarkson et al., 2017). These findings suggest that kisspeptin arcuate nucleus activity is acting on GnRH neurons to govern the pulsatile release and subsequently regulate LH activity. But what happens to the GnRH pulsatile activity in the absence of kisspeptin? During development, normal wild-type female mice reach puberty at approximately 30 days of age, indicated by vaginal opening. In mice that had either the KISS or GPR54 gene deleted, female mice showed vaginal opening after approximately 50 days of age (i.e., delayed puberty). However, both male and female mice had very low gonad weight and were incapable of any successful pregnancy, indicating that the GnRH neurosecretory system was not operant as in controls (Lapatto et al., 2007). Yet, in these mice, the pituitary gland still released LH, and the temporary inhibition of arcuate nucleus kisspeptin neuronal activity caused an overall reduction in LH pulsatile concentrations and a decrease in pulse amplitude. It is now well established that a small population of arcuate nucleus kisspeptin neurons that co-express neurokinin B and dynorphin, termed KNDy neurons, provide negative input to GnRH cell bodies and dendrites and decrease pulsatile activity (Moore et al., 2018). KNDy cells also express estrogen receptors and provide one of the critical links for elucidating how ovarian steroids provide negative feedback for the homeostatic regulation of reproductive physiology in both males and females.

But in all the experiments that assess the functional role of kisspeptin, none have shown a complete loss of LH pulsatile activity. Indeed, several other neuromodulators are also involved in the regulation of GnRH pulsatility. The predominant inhibitory neurotransmitter, glutamic acid decarboxylase (GABA), maintains constant contact with multiple GnRH neurons and functions to "gate" information communicated to the cells from other modulators (Leranth et al., 1985). Norepinephrine cells located in the midbrain send long projections into the preoptic area of the hypothalamus and provide both stimulatory and inhibitory input to GnRH cells with direct impacts on pulsatile activity (Taliesnik and Sawyer, 1986). Similar to all neuropeptides that exhibit pulsatile activity, GnRH release can be modulated by many neuromodulators, and a small list includes catecholamines, glutamate, histamine, and serotonin, as well as neuropeptides such as corticotropin-releasing hormone, neuropeptide Y, substance P, and vasoactive intestinal polypeptide. An excellent compilation of information on the GnRH pulsatile activity and regulation by other neuronal systems can be found in the book *GnRH: the master molecule of reproduction* (2002) by Andrea Gore.

But as we discussed above, one of the key features of central pattern generators is the ability to maintain rhythmic activity independent of external (i.e., other neuronal) input. For the multiple examples of other neuromodulatory systems outlined here, there is a significant impact on the frequency and/or amplitude of the GnRH pulse generator, yet at a fundamental level, GnRH pulsatility persists. The alternative explanation then is that GnRH neurons also themselves possess a central pattern generator. If so, GnRH neurons must themselves possess not just intrinsic pulsatile but surge properties as well.

GnRH cells are central pattern generators for programmed regulation of female cycles

The question of whether GnRH neurons themselves possess intrinsic activity can only be addressed using GnRH cell populations that are isolated from all other neuromodulatory input. Professor Pamela Mellon, at the University of California School of Medicine, conducted an ingenious study with colleagues when she worked at the Salk Institute, in which she extracted GnRH neurons from the mouse brain (Mellon et al., 1990). The isolation of single GnRH neurons was a major milestone in understanding how the brain controls reproduction. These GnRH cells, now referred to as GT1 cells, are immortalized, meaning the cells will continue to multiply indefinitely in cell culture. In cell culture dishes used for neuronal experiments, GT1 cells have the same morphology and function in a manner like GnRH neurons in the hypothalamus. Work in Richard Weiner's laboratory at the University of California, San Francisco revealed that GT1 cells maintain an "inherent" pulsatility (Martinez de la Escalera et al., 1992). The ability of GT1 cells to release GnRH in a rhythmic manner demonstrates that there is an intrinsic characteristic of these neurons. What is particularly unusual is that cultured cells with established cell-to-cell contacts show a synchronization in GnRH release! GnRH neurons form a network that communicates with each other via gap junctions, and this serves to increase activity across the population and also increase pulsatile coordination. Therefore, other neuromodulators simply increase the amplitude of the GnRH pulsatile activity and are not necessarily the only components that drive pulse activity. But what about the programmed rheostatic balance that is maintained over the entire reproductive cycle? The answer to this lies in how the GnRH cells maintain "surge" activity across long-term timescales.

The GnRH pulse generator oscillates over relatively short timescales that are species-specific and range from minutes to several hours. What are the similarities and differences for the control of the GnRH pulse generator versus the GnRH surge generator? The evidence indicates that both hormonal feedback in the form of positive stimulation by estrogen and a circadian pacemaker collectively function to time the GnRH surge and ultimately ovulation in rodents. Again, the vast majority of the research has been conducted in female mice, due to the ability to employ non-invasive measures to track the estrous cycle and then use precise timing to couple studies into the robust changes in hormone concentrations and neuroendocrine plasticity. The key advantage of mouse models is the capability to induce selective genetic mutations to identify key genes that can dissociate pulsatile from surge activity.

Role of circadian time in the control of female reproductive cycles

We have already discussed the evidence showing that programmed circadian rheostasis explains the daily change in homeostatic control of physiological systems. Reproductive hormones also exhibit marked age-dependent daily rhythms in blood concentrations. In male humans, blood testosterone concentrations in young men vary from approximately 8 ng/ml to 5 ng/ml with peak levels early in the morning (i.e., 0800h) and a trough in concentrations in the early evening (2000h). As men age there is an overall decrease in plasma testosterone concentrations and a loss in the robustness of the daily rhythms (Bremer et al., 1983). Similarly, the blood concentrations of estradiol, a specific molecular structure of estrogen, also show daily rhythms with peak concentrations in the morning (0600h) and a trough during the early evening (1800h) (Bao et al., 2003). Circadian rhythmicity is observed across the entire neuroendocrine-pituitary-gonadal axis with bidirectional relations between cell-autonomous circadian clock genes and gonadal steroid (i.e., estrogen/testosterone) receptors in multiple brain regions and peripheral tissues. For an excellent overview of circadian timing in male and female reproductive systems and the implications of disrupted daily rhythms for disease, see the comprehensive review by Matthew Bailey and Rae Silver (2014). The role of circadian clock timing in physiological systems has already been covered in sufficient detail in Chapter 3.

In rodents, disruption of the circadian pacemaker nucleus SCN eliminates the successful daily timing of the LH surge (Gray et al., 1978). The neuropeptide vasoactive intestinal polypeptide (VIP) is expressed in cells located in the SCN and projects to a subset of GnRH neurons. VIP expressing SCN cells then act on VIP2 receptors to regulate the daily timing of the surge activity (Van der Beek et al., 1997). In the GT1-7 cells, the circadian clock genes show robust daily rhythms in expression and may function to alter the sensitivity of GnRH release to the daily timing in exposure to both kisspeptin and VIP (Zhao and Kriegsfeld, 2009). In other animals, such as sheep, the role of SCN in the timing of the GnRH surge is less important. In ewes, the GnRH surge generator is dependent on the amount of time after an initial "priming" step driven by estrogen (Goodman et al., 1981). These findings raise the potential that GnRH neurons independently time the surge generator in response to ovarian hormone priming that is later fine-tuned by cell-specific circadian clock genes and/or VIP input from SCN.

Studies in female sheep and mice have provided the most complete hormonal, molecular, and neuronal circuit for the regulation of GnRH surge and the timing of ovulation (Figure 4.4). Here, kisspeptin neurons are again central in the regulation of GnRH activity in mammals. Another kisspeptin population–located in the preoptic area/anteroventro-periventricular region in the anterior hypothalamus–expresses estrogen receptors and mediates positive feedback effects to GnRH cell bodies. The daily programmed rheostatic timing of GnRH activity has direct and indirect pathways from the circadian pacemaker, the SCN. While there is a VIP projection directly from the SCN that innervates with many GnRH neurons in both male and females, it surprisingly does not impact GnRH surge activity (Piet et a., 2016). So, despite daily input to the GnRH systems from the SCN there is no clear direct functional link that conveys time-of-day information. An alternative indirect pathway via the AvPv kisspeptin population serves to time and amplify the GnRH surge. Another neuropeptide system, vasopressin (AVP) expressing cells in the SCN project to the AvPv kisspeptin neurons and provide the daily programmed rheostatic drive that times GnRH surge activity (Williams et al., 2011). This additional higher order pathway includes a discrete set of genetic, cellular, and neuronal circuit systems that is entirely separate from the short-term homeostatic regulation of reproductive physiology.

Even though kisspeptin is a critical gene for timing puberty and fertility in most mammals, it is unlikely to be an evolutionary conserved central pattern generator for the regulation of female reproductive cycles. Instead, kisspeptin provides an additional level of organization in the hypothalamus to integrate multiple environmental and internal signals and serves to amplify the GnRH surge signal into the hypophyseal portal (Kriegsfeld and

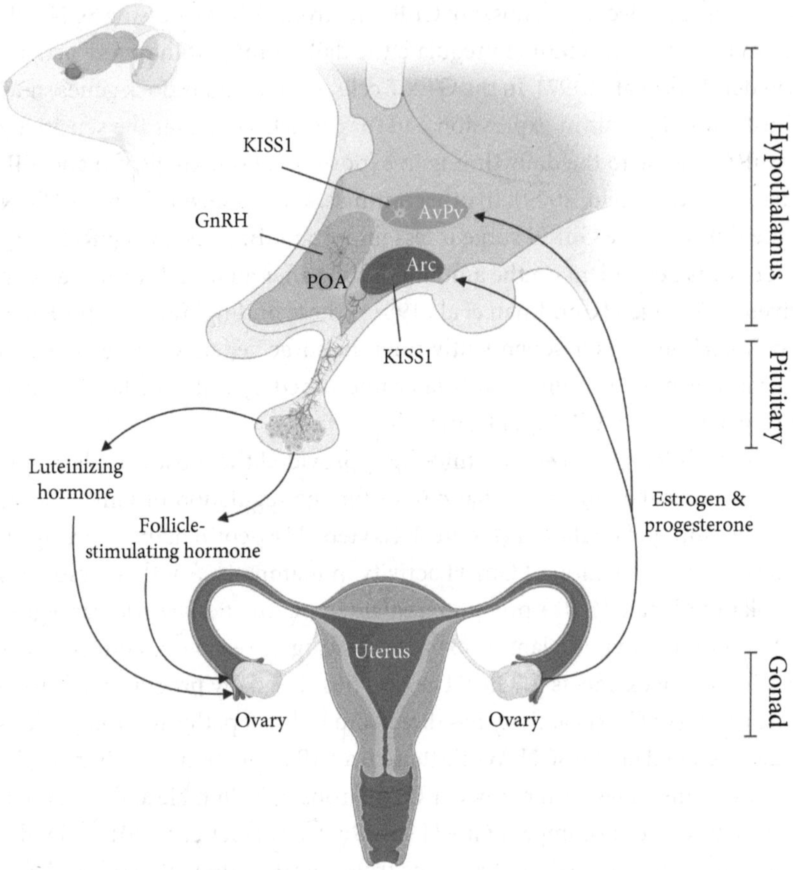

Figure 4.4 Programmed rheostatic and homeostatic control of reproductive physiology.
Gonadotropin-releasing hormone (GnRH) neurons in the preoptic area (POA) stimulate the release of gonadotropins, luteinizing hormone and follicle stimulating hormone, from the pituitary gland. Gonadotropins signal via receptors located in the ovary to produce gonadal sex steroids (i.e., estrogen and progesterone). These ovarian steroids act in the hypothalamus to trigger negative feedback (via Arc) and positive feedback (via AvPv) to kisspeptin (*Kiss1*). Programmed rheostasis involves the coordination of negative and positive feedback from *Kiss1* to GnRH neurons, as well as central pattern generators localized in GnRH neurons.

Williams, 2012). What is the evidence supports such a proposition? Let us turn now to another amniote: birds.

Long-term regulated changes of reproductive cycles in female birds

Although significantly less studied, there are large, programmed rhythms in reproductive hormones in birds. Egg laying usually occurs daily and is dependent on day length and age. Most birds, except a few such as the emu, are long-day breeders and engage in egg laying during summer periods. The daily timing of egg laying shows species-specificity; for example, chickens and duck laying during the dusk phase, while Japanese quail during the dawn phase (Cunningham et al., 1984). As in mammals, ovulation is preceded by a massive increase in LH release from the anterior pituitary gland (Cunningham et al., 1984). In turkey hens, there is a robust LH surge that shows a higher frequency in excellent versus poor egg laying birds (Liu et al., 2001). In Japanese quail, the LH surge occurs 10–6 hours before ovulation and is primarily driven by hormones secreted by the largest follicle (i.e., progesterone) (Doi et al., 1980). The LH surge in birds is also driven by GnRH activity (Johnson et al., 1984). But, in stark contrast to the mammalian system, birds have neither the kisspeptin gene nor the receptor for kisspeptin (i.e., GPR54). The lack of these two genes signifies that the kisspeptin system is not wholly responsible for programmed rheostasis in female reproductive cycles.

In birds, the GnRH system is the sole integrator of hormonal and neuronal cues and times reproductive activity (Stevenson et al., 2012). Most non-domesticated birds use the annual change in environmental cues such as daylength to time GnRH activity and egg laying to an optimal period of the year (MacDougall-Shackleton et al., 2009). In the absence of kisspeptin, the GnRH neuronal system functions as both an error detector and a controller for maintaining stability in reproductive physiology. How the GnRH system in birds differentiates both homeostatic and rheostatic systems in still a mystery. But current evidence indicates that the circadian system, such as the SCN, and circadian clock genes are involved in timing daily rhythms in egg-laying in addition to intrinsic GnRH pulsatile and surge generation (Follett, 2015).

Theoretical framework for the programmed rheostatic control of reproductive cycles

Now that we have reviewed the genetic, hormonal, and neuronal pathways that are similar and different between reproductive homeostasis and rheostasis, we will discuss how to develop a model framework for rheostasis. A theoretical framework for the programmed rheostatic control of long-term stability in the female reproductive cycles can be achieved by expanding upon the model outlined in Chapter 3 (Figure 4.5). In this programmed

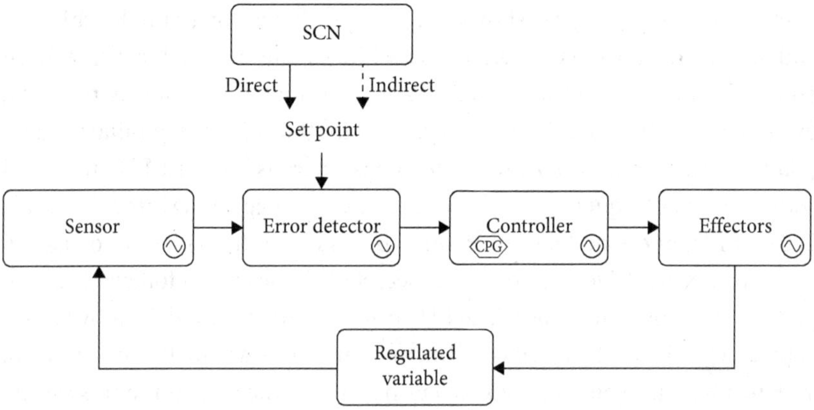

Figure 4.5 Schematic model of programmed rheostasis of female reproductive cycles.
Rheostatic control of female reproductive cycles has distinct components that can be separated from the homeostatic regulation of reproductive physiology. The major difference is the timing of positive feedback. In this model, the main components that are different reside in the error detector and controller and in how input from the suprachiasmatic nucleus (SCN) times daily variation in GnRH surge activity. The two programmed factors that contribute to programmed rheostasis in female reproductive cycles include the SCN and neuropeptide central pattern generators (CPG). In mammals, kisspeptin cells located in the anteroventral-periventricular and preoptic area nucleus function as an error detector and integrate hormonal signals (e.g., estrogen) and the circadian timing mechanism directly, indirectly, and by circadian clock genes indicated by rhythmic symbol. GnRH cells in the preoptic area function as the controller and receive input from error detector signals, but also can drive programmed changes in set point values via intrinsic CPG activity. In birds, the GnRH system functions as both error detector and controller.

reproductive rheostat, we will use the blood concentrations of ovarian steroids as the regulated variable. In this scenario, the sensor must include receptors that detect the circulating levels of estrogen (and progesterone). The cellular integration of estrogen-receptor complex signaling primarily occurs in the arcuate nucleus but may also include estrogen-dependent signaling of the reproductive cycle conveyed by other hypothalamic (e.g., ventromedial nucleus) and extra-hypothalamic nuclei (e.g., periaqueductal gray). In rodents and humans, the kisspeptin neurons in the anteroventral periventricular nucleus and preoptic area act as the error detector. These neurons possess circadian clock genes that provide daily time and in conjunction with SCN input, provide the long-term timing input to the regulated set point. These kisspeptin neuronal networks project to the GnRH neurons and here serve to facilitate the intrinsic pulse generator output. These GnRH cells form the controller and provide the species-specific regulation of the LH surge. It is important to emphasize that despite the intrinsic activity of central pattern generators (i.e., GnRH cells) being frequency-dependent, the oscillations do not establish the period of either estrous or menstrual cycles. The timing of estrous and menstrual cycles comes from the respective lengths of luteal and follicular phases in the ovary. In the absence of fertilization and pregnancy during the luteal phase, no corpus luteum forms in rodents or degrades in sheep and primates. Then, a new follicular stage is established, and other follicle(s) develop. So, the GnRH cells themselves do not maintain long-term regulated cyclicity (e.g., 28-day menstrual cycles). In other vertebrates, such as birds, it is likely that GnRH neurons have retained the central pattern generator as the predominant intrinsic timing mechanism, and this acts as both the error detector and controller. For mammals and birds, the daily timing of GnRH surge activity is dependent on a functional circadian pacemaker and AVP cells localized in the SCN add another layer of organization for the control of set points of regulated variables across the female reproductive cycle. The final pathway for the model is the GnRH output to the anterior pituitary gland in which the gonadotropins act as the effector to drive steroidogenesis and gametogenesis.

Summary

In this chapter, the female reproductive cycle was used to demonstrate that programmed changes in physiology can occur over time scales longer than a day. The short-term homeostatic regulation of regulated reproductive variables, such as gonadotropins and gonadal steroids (i.e., estrogen), maintains stability using negative feedback that includes kisspeptin gene expression in the arcuate nucleus of the hypothalamus, or direct feedback to GnRH cells in the case of birds (Stevenson and Ball, 2009). It was proposed that long-term programmed reproductive rheostasis used a different set of cells (i.e., anteroventral periventricular kisspeptin) and neuronal circuitry that is species-specific. The concept of and evidence for central pattern generators maintaining the programmed rhythmic activity in neuropeptide activity and GnRH cells having an intrinsic nature to time pulsatile and surge activity was highlighted. This property is common across neuropeptides and can be generated in the absence of environmental input. Neuromodulators such as kisspeptin, norepinephrine, and GABA can directly fine-tune the programmed release, and predominantly act to increase GnRH amplitude. In addition, positive feedback from gonadal hormones serves to increase the intensity of the GnRH and LH surge. Therefore, programmed reproductive rheostasis is established by (1) positive feedback from ovarian estrogen triggered by the gradual rise in estrogen concentrations, which reflects follicular development and follicular granulosa cell activity and (2) a switch from the arcuate nucleus kisspeptin regulation to anteroventricular periventricular kisspeptin signaling. Now that programmed rheostasis has been outlined for daily, circaquadridian (i.e., 4-day estrous), and circalunar (i.e., 28-day menstrual) cycles, let's turn to significantly longer programmed rhythms, the yearly or circannual cycle.

5

A seasonally programmed energy rheostat

> I have to go to bed and see, the birds still hopping on the tree, or hear the grown-up people's feet, still going past me in the street.
>
> Robert Louis Stevenson, A Child's Garden of Verses, 1885

One of the most dramatic changes in the environment is the annual cycle in temperate zone day lengths. The North Pole will experience approximately 187 and 162 days of constant daylight or darkness, respectively. In Aberdeen, Scotland, the summer solstice witnesses almost 18 hours of sunlight compared to the winter solstice with a daylength of only about 7 hours. During the summer nights, it was not uncommon to hear the "grown-ups" referenced by Robert Louis Stevenson walking down Clarks Lane, the street in which we lived, during the early hours of the morning. Humans have long used the annual change in daylength as a predictive cue to time seasonal agriculture. Aubrey Burl and Edward Piper (1979) in *Rings of Stone: The Prehistoric Stone Circles of Britain and Ireland* identified and characterized the vast number and distribution of late Neolithic monuments, many aligned with the sun or moon. The county of Aberdeenshire has over 100 known stone circles, many of which are *recumbent* stone circles. The simplicity in the design of recumbent stone circles is quite elegant. On the southwest side of the circle, the recumbent stone is positioned laying sideways with two *flankers* erected on either side of it. Another series of stones were positioned with remarkable precision to create the circumference of the circle with the smallest stone opposite to the *recumbent*. A setting full moon during the deep dark of winter is reliably captured by the recumbent stone and provides an annual marker to start tracking the time (or phase) of the year. An astute farmer could then simply count the number of subsequent full moons and mark each lunar event on the stones that form the stone circle. In this manner, the entire year could be tracked by counting the twelve full moons—a rather

On Rheostasis. Tyler John Stevenson, Oxford University Press. © Oxford University Press 2024.
DOI: 10.1093/oso/9780197665572.003.0005

ingenious system for a culture that required precise knowledge of the time of year to optimize the planting of crops.

Similar stone monuments can be found in the Southern Hemisphere. The Chankillo Towers consist of thirteen cuboidal stone pillars along a hill range in the Casma-Sechin basin of Ancash Department in Peru. The line of stone pillars is aligned north to south and functions as a "solar calendar." From relatively distant observation points, the community could easily track the annual change in the position of the sun. During the June solstice, sunrise would be captured by the northern pillar, and the December solstice would be captured by the southern pillar (Ghezzi and Ruggles, 2007). Other stones positioned between the end pillars could reliably identify the two equinoxes and the (anti)zenith sunrise.

Neolithic stone circles were also used by relatively equatorial human populations. In the Nabta Playa depression in southern Egypt, there is a stone circle that could identify the summer solstice sunrise and could serve as an annual calendar to predict monsoon seasons (Malville et al., 1998). These are just three examples of very different annual clocks used by early humans, mostly agricultural communities, to not only identify the current annual phase but also anticipate the directional change in seasons. Indeed, ancient civilizations had deities such as Persephone, that symbolized the annual spring growth of vegetation after a prolonged period of dormancy.

Humans now use a range of technological advances to track local and global climates, with increased agricultural yields as one positive outcome. Most animals living in the temperate zone rely on seasonal changes in nutrient availability to develop sufficient stores of energy in the form of fat cells to survive seasons of drought and poor food availability. During those nutrient poor months, animal adaptations consist of dormancy such as diapause or hibernation, or energetically demanding behavior such as migration. For most, an annual interval timer such as those described in Chapter 1 is used to anticipate future environments. It is not uncommon for plants and animals to have evolved long-term interval timers that govern regulated changes in energy stability across phases of nutrient abundance and paucity to maintain animal fitness across the year, and across multiple years for long-lived species.

One simple overview of energy stability is the coordinated regulation of food intake (energy input) and expenditure (energy outflow). The canonical equation to illustrate energy stability is derived from a balance between *energy intake (food + fluids)* with *energy expended (physical activity levels, resting energy expenditure, thermal effect of food*). Energy homeostasis, or the steady state between energy intake and energy expenditure, is required for short-term or acute changes in energy stores lasting over periods of hours. For the purposes of this chapter, energy homeostasis involves the short-term changes that are observed within a day, whereas long-term, programmed energy rheostasis is required for regulated changes in energy balance over periods of days and months. As we will discuss, energy homeostasis involves the coordination of multiple processes with contributions from genetic, physiological, and behavioral systems. In this chapter, the complex nature of energy homeostasis will be discussed first, followed by energy rheostasis using the Siberian hamster as a model. As in Chapters 3 and 4, the genetic, molecular, and cellular processes and the circuits that differentiate homeostatic and rheostatic processes will be discussed to highlight clear differences in physiological control.

Physiological basis of energy homeostasis

Multiple different regulated variables could be selected to illustrate energy stability. For most of us, our body weight is one variable that is quite easy to understand and is likely the most common variable people try to control. A key measurement of energy stability that is used to monitor human "health" is the body mass index (BMI). The BMI is a relative score that combines weight (in kilograms) and height (in meters). It is a rather easy calculation and can be determined by dividing an individual's body weight by the square root of the height and is expressed as kg/m^2. The BMI index is a useful measure for understanding human population health. The World Health Organization (WHO) has developed a chart for international use, and it provides estimated cut-off values between the different BMI categories. For example, the average normal BMI for a human ranges between 18.5 and 25. Humans that are considered underweight have a BMI <18.5; overweight individuals span 25–30; and obese individuals have a BMI >30. Although sugar levels are generally not associated with BMI values, the amount of fat (also referred to as adipose) is the predominant physical feature and contributor that drives overweight and obesity diagnoses.

One of the main sources of energy that contributes to an individual or animal's BMI comes from a sugar that circulates in the form of blood glucose and is derived from several different food resources such as fruit.

Diabetes is a condition in which there are constant, high levels of sugar found circulating in the blood. Historically, people were identified as diabetic due to the sweetness in the smell of their urine. Until the early 1900, it was generally assumed that diabetes was a kidney condition and the inability to reduce blood glucose levels was due to an ineffective "renal" hormone. One of the most important discoveries of the twentieth century was achieved by Sir Frederick Banting (1891–1941), a Canadian physician who spent most of his life in London, Ontario, and Toronto, Ontario.[1] He is the youngest Nobel laureate in the discipline of Physiology or Medicine, having received the award in 1923 at the age of 31 (Bliss, 1992). This honor was due to the critical breakthrough of isolating insulin from a key cell population in the pancreas, the Islets of Langerhans. Individuals that suffer from diabetes either do not make insulin and blood glucose levels become too high (Type 1 diabetes) or insulin resistance occurs and generally results in high blood glucose due to poor insulin regulation (Type 2 diabetes). Insulin secretion from the pancreas is critical for moving glucose out of the blood and into cells that break down the sugar molecule to produce energy in the form of adenosine phosphate. Increased circulating insulin also signals to the brain there are higher levels of blood glucose (Figure 5.1). The increase in activity of insulin initiates a "satiety" signal or the feeling of being "full" and results in reduced food intake.

During the 1940s it was becoming apparent that in addition to psychological (i.e., feeding) and physiological (i.e., hormone) regulators of energy homeostasis, a genetic component must be involved. An obese phenotype associated with hyperphagia (i.e., overeating), decreased energy expenditure, and Type-2 diabetes was found in a colony of mice kept in captivity. The obese phenotype was a recessive mutant, meaning the function of the gene was lost, and some members of the colony (approximately 25%) were observed to have excessive weight gain (Ingalls et al., 1950). In 1994, Zhang and colleagues sequenced the obese gene (*ob*) and showed that when mutated the result in mice was significant weight gain. The obese gene was called leptin and found to be produced by white adipose tissue and secreted into the circulatory system. Crucially, leptin is produced in proportion to the amount of adipose tissue, so the more fat in the body, the more the hormone is secreted into the circulatory system. After food is digested, excess carbohydrates, protein, and glucose above those required for immediate functions are stored in the form of fat and provide another "satiety" signal to the brain that serves to stop food intake (Figure 5.1).

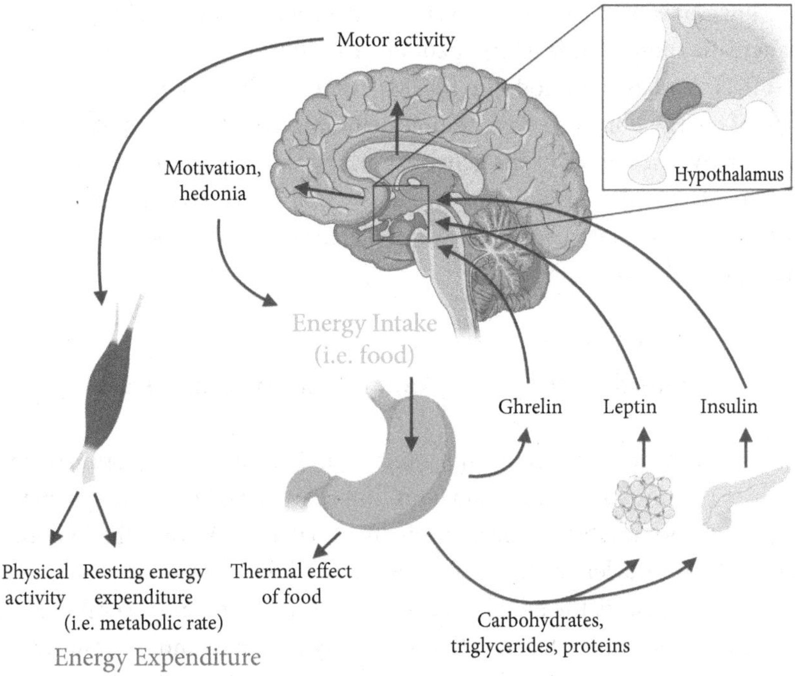

Figure 5.1 Physiological pathways involved in energy homeostasis. Energy homeostasis involves the balance between energy intake (i.e., carbohydrates, protein, triglycerides) and energy expenditure (i.e., physical activity, resting energy expenditure, thermal effect of food). Energy is readily stored in peripheral tissues (e.g., adipose/fat). The hormone leptin is produced in proportion to the amount of stored adipose cells and provides an internal signal of excess energy. In addition to leptin, insulin, and ghrelin are involved in signaling satiety and hunger.

One of the quickest physiological means to stimulate food intake is the production and secretion of the stomach hormone ghrelin into the circulatory system. Ghrelin was discovered in 1999; it is released from an empty stomach and plays a major role in the regulation of food intake (Kojima et al., 1999). Ghrelin has central actions in the brain because injections directly into the hypothalamus stimulate food intake and increased body weight gain in rats (Nakazato et al., 2001). Inhibition of ghrelin using pharmacological agents results in decreased food intake and body weight (Tschop et al., 2000). Remarkably, ghrelin is the only known *orexigenic* (meaning appetite stimulant) hormone produced by the gut. A constellation of other *anorexigenic* gut hormones have been discovered and include cholecystokinin (CCK), pancreatic polypeptide (PP), peptide YY (PYY), glucagon-like peptide (GLP-1), and oxyntomodulin (OXM). Indeed, the physiological regulation of energy homeostasis is complex, and the scientific discipline has had many new discoveries in the genetic, hormonal, and neural control of food intake. The aim here is to focus on the few well-characterized physiological components of energy homeostasis: a satiated (i.e., leptin, insulin) and fasted (i.e., ghrelin) state.

Hypothalamic regulation of energy homeostasis

Circulating satiety and hunger hormonal signals act in multiple brain regions such as the hindbrain, midbrain, and limbic systems, but the key structure for energy homeostasis resides in the hypothalamus. Within the hypothalamus several nuclei are implicated in energy homeostasis and the central integratory locus is the arcuate nucleus because it contains several neurons and glial cells (e.g., tanycytes) that have easy access to and respond to circulating hormones involved in energy homeostasis. Crucially, the receptors of insulin (Marks et al., 1990), leptin (Elmquist et al., 1998), and ghrelin (Zigman et al., 2005) show abundant expression in the arcuate nucleus in mammals and in many instances are in the same neurons. Using mice, studies have revealed that the conditional mutagenesis of insulin receptors results in mild metabolic abnormalities including substantial weight gain and the development of diabetes (Kitamura et al., 2003). Mutation of the leptin receptor (*ObR*) results in significantly overweight mice that develop obesity-like conditions such as elevated leptin and glucose (Cohen et al., 2001). Conversely, the selective deletion of the ghrelin receptor results in mice that eat less food, accumulate less body weight, store fewer calories, and have as their primary energy substrate fat deposits from adipose tissue (Zigman et al., 2005).

A series of discrete, interconnected neuronal populations have been discovered in the arcuate nucleus that are critical for the integration of hormonal signals for energy homeostasis. Specifically, two neuronal populations that both express two neuropeptides have been identified and shown to stimulate or inhibit food intake. One neuronal population includes cells that express agouti-related peptide (Agrp) and neuropeptide Y (Npy). Agrp and Npy cells are responsive to insulin, leptin, and ghrelin hormonal signals. The other neural populations are proopiomelanocortin (Pomc) and cocaine-related transcript (Cart) expressing cells that also integrate hormonal signals, but evidence to date suggests these cells only respond to insulin and leptin (Figure 5.2).

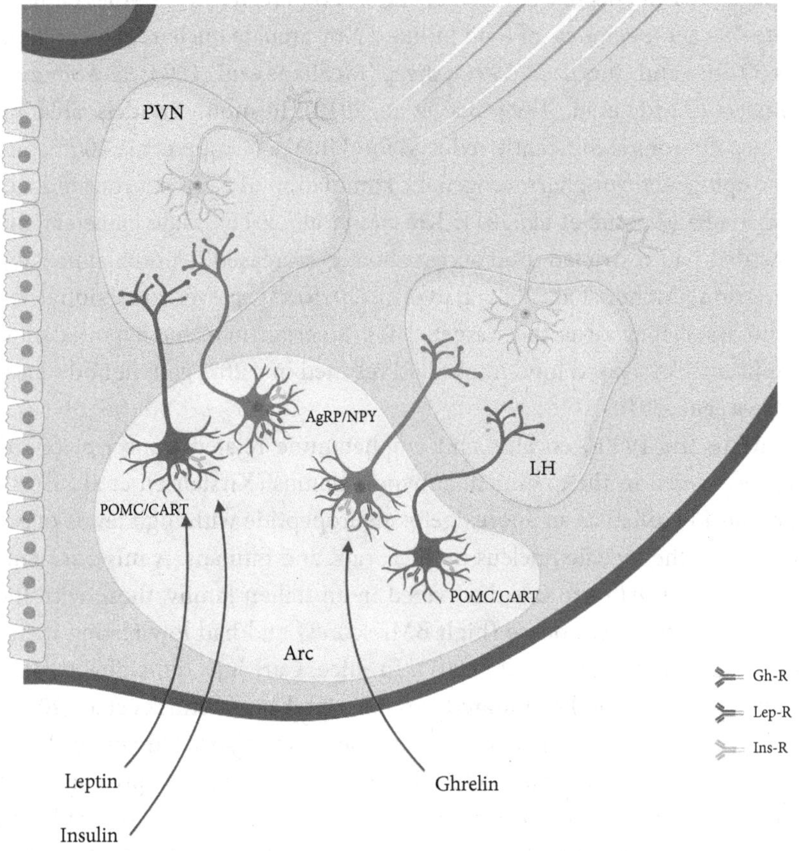

Figure 5.2 Hypothalamic neuronal circuit for energy homeostasis. Hormones that signal satiety (i.e., leptin, insulin) and hunger (i.e., ghrelin) bind to receptors located on neuropeptides in the arcuate nucleus (Arc). Two types of neuroendocrine cells have been classified as orexigenic and anorexigenic. Leptin and insulin decrease orexigenic agouti-related peptide (*Agrp*) and neuropeptide Y (*Npy*) and increase anorexigenic proopiomelanocortin (*Pomc*) and cocaine- and amphetamine-related transcript (*Cart*). Ghrelin increases *Npy* and *Agrp* and reduces *Pomc*. Leptin and insulin act to inhibit while ghrelin increases appetite.

Npy is a major orexigenic neuropeptide expressed in the arcuate nucleus and stimulates food intake by acting in the paraventricular (PVN) and ventromedial nuclei (VMN) (Beck, 2006). Npy injections directly into the hypothalamus induce potent increases in acute (e.g., 6–8hr) food intake in multiple species including mice (Morley et al., 1987), rats (Clark et al., 1984), sheep (Miner et al., 1989), and chickens (Kuenzel et al., 1987) and increases adipogenesis and decreases thermogenesis. Food restriction studies demonstrate that acute removal of food induced Npy arcuate nucleus expression in rats (O'Shea and Gundlach, 1991), sheep (McShane et al., 1993), and Siberian hamsters (Reddy et al., 1999; Bao et al., 2019). In mouse models, ablation of Agrp neurons significantly reduced food intake (Gropp et al., 2005), and acute optogenetic or pharmacogenetic stimulation of Agrp neurons induces food intake (Aponte et al., 2011; Krashes et al., 2011). Male hamsters and rams that had restricted food access showed increased hypothalamic *Agrp* expression (Archer et al., 2004; Bao et al., 2019). Agrp overexpression using adeno-associated virus that targeted the Siberian hamster hypothalamus significantly increased food intake and resulted in a 20% gain in body mass (Jethwa et al., 2010).

Late in the 1990s, cocaine and amphetamine related transcript (Cart) was discovered in the mammalian hypothalamus (Kirstensen et al., 1998). Cart was identified as an anorexigenic neuropeptide with high levels of expression in the arcuate nucleus in mice, rats, and humans. A missense mutation in the *Cart* gene was discovered in an Italian family, those with the genomic change were obese (high BMI scores) and had low resting metabolic rates (del Giudice et al., 2001). In mice, Cart injections directly into the third ventricle of the brain reduces food intake (Maletinska et al., 2008). A reduction in circulating leptin concentrations is associated with reduced *Cart* expression (Xiao et al., 2005). In mice, an exogenous leptin injection produced an increased *Cart* expression (Kristensen et al., 1998), and leptin receptors are expressed on Cart-expressing cells (Elias et al., 2001). Several studies have indicated that *Cart* is acutely regulated by energetic state (Yeo and Heisler, 2012), and food deprivation significantly decreased *Cart* expression (Kirstensen et al., 1998).

Unlike Npy, Agrp, and Cart, the neuropeptide Pomc gene includes a family of pre-prohormones that individually lead to the synthesis of melanocyte stimulating hormone (Msh), adrenocorticotropic releasing hormone (Acth), β-lipotropin hormone, and β-endorphin (Figure 5.2). Pomc neurons are expressed in the arcuate nucleus and the nucleus of the solitary tract in the brainstem (Cone, 2005; Yeo and Heisler, 2012). The combination of the melanocyte peptides (αMsh and βMsh) and associated melanocortin receptors (Mcr) form the "melanocortin system," which is fundamental for the control of food intake. Professor Sadaf Farooqi and Stephen O'Rahilly at the Wellcome-MRC Institute of Metabolic Science at the University of Cambridge have made a series of discoveries on the role of genes in the melanocortin system for our understanding of obesity in humans. This group examined the Pomc gene sequences from hundreds of UK children with severe obesity. Of the genes identified, one was located in the αMsh sequence and two were located in the βMsh sequence. Mutations in βMsh reduce the effectiveness of the protein in binding to the melanocortin receptor type 4 (Mc4r) (Lee et al., 2006). This lack of activation of Mc4r results in obese/overweight children. Of notable importance is the difference between human populations and mice, as αMsh mutations cause more severe obese states than the *βMsh* form. Mice that harbor a deletion in the Pomc-derived peptides are obese and when treated with Pomc-derived analogs (e.g., αMsh) show a reduction in weight after 2 weeks' treatment (Yaswen et al., 1999).

Five melanocortin receptors have been identified in mammals, and the Mc4r gene is the dominant receptor for controlling energy homeostasis. In 1998, Farooqi's group identified a family that carried a mutation in the Mc4r gene that was associated with obesity (Yeo et al., 1998). Simultaneously, another research group at the Institut Pasteur de Lille also identified a family that had another mutation in the Mc4r gene and were morbidly obese with significantly elevated leptin levels (Vaisse et al., 1998). Binding of βMsh to the Mc4r receptor reduces food intake. Conversely, Agrp also binds to Mc4r but has the complete opposite effect, and instead acts as an antagonist to stimulate food intake (Ollmann et al., 1997). These findings indicate that the Pomc and Agrp peptides in the arcuate nucleus integrate hormonal signals from leptin and insulin and converge on cells that express Mc4r to regulate energy stability.

Framework for energy homeostasis

Altogether, the hormonal satiety signals derived from leptin and insulin lead to a reduction in the Agrp and Npy signaling and increase Pomc and Cart expression, which in turn reduce food intake. Ghrelin binds to the ghrelin receptor on Agrp and Npy cells and directly increases activity leading to food intake. These data indicate that the central node for the integration of energy homeostasis resides in the arcuate nucleus. The output signals from the arcuate are complex and involve a range of other brain regions. Pomc/Cart projections to the PVN act via Mc4r to induce anorexigenic effects on energy homeostasis. Agrp/Npy cells also project to neurons in the PVN that express Mc4r and competitively antagonize the receptor to induce orexigenic effects on food intake. Other brain regions including the parabrachial nucleus in the dorsal vagal complex (DVC) in the midbrain, and the intermediolateral column (IML) of the spinal cord are targets of arcuate neuropeptide signaling involved in energy homeostasis (Sohn et al., 2013). Moreover, the Mc4r expressing neurons in the PVN project to the DVC and parabrachial nucleus. There, the DVC and IML regulate energy expenditure via autonomic preganglionic neurons that express Mc4r to regulate food intake.

Using the hormonal, neuronal and neural circuits described above, we can create a model for energy homeostasis (Figure 5.3). Here, we will use the WHO definition of healthy BMI as the working set point. As stated above, leptin and insulin are the two predominant hormones secreted by adipose

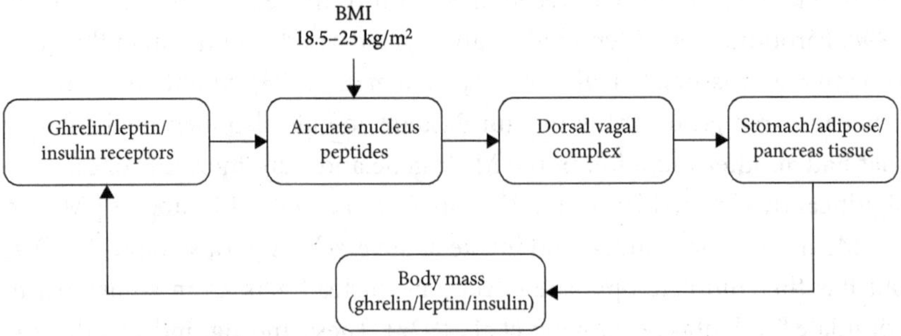

Figure 5.3 Schematic overview of energy homeostasis.
A simplistic model of energy homeostasis centers on the maintenance of relatively constant body mass. Leptin (and to an extent insulin) are synthesized in proportion to the amount of body fat and provide a physiological readout of body mass. Leptin/insulin receptors in the arcuate nucleus function to detect any change in circulating hormone levels in the blood. An interconnected network of neuropeptides in the arcuate nucleus integrates hormonal signals and compares with an individual's BMI set point. The arcuate nucleus projects directly to the dorsal vagal complex and indirectly via the PVN to control food intake, physical activity, resting expenditure, and the thermic effect of food. Changes in energy expenditure impacts peripheral adipose/pancreatic function to reduce/increase body mass and insulin/leptin secretion.

tissue and the pancreatic islets and are produced in proportion to the size of the tissue, providing a reliable indicator of BMI. Receptors for both leptin and insulin are in the arcuate nucleus and serve as sensors of the acute changes in circulating hormone levels. These receptors are located on two classes of neuropeptides (i.e., Pomc/Cart and Agrp/Npy), which signal to increase food intake (orexigenic) or reduce food intake (anorexigenic) behavior. The neuropeptides act as sensors to maintain a body mass that is approximately 18.5–25kg/m^2 and project to multiple downstream brain regions that express Mc4r to coordinate food intake and energy expenditure. The dorsal vagal complex in particular acts as the "controller" to integrate the orexigenic βMsh in human (αMsh in rodents) and anorexigenic Agrp to regulate energy expenditure. Increased signaling of energy expenditure by the dorsal vagal complex is passed via sympathetic connections to trigger the breakdown of adipose tissue (i.e., lipolysis) and the secretion of leptin from adipocytes and insulin from the pancreas.

Seasonal programmed energy rheostasis

These systems are critical for the short-term homeostatic regulation of energy homeostasis across the animal kingdom. But mammals, birds, amphibians, reptiles, and fish also show predictable long-term rhythms in energy balance. The annual change in food intake and energy expenditure does not conform to the simple energy homeostasis model yet fits remarkably well with the concept of programmed rheostasis. Annual cycles in energy rheostasis are a critical adaptation of animals to survive seasonally challenging environments and essential to ensure the successful timing of major life-history events such as migration and hibernation.

First, it is worthwhile to emphasize that humans are not excluded from programmed long-term physiological changes in energy rheostasis. The best evidence to support population rhythms in human energy balance can easily be obtained from using massive online data repositories from hundreds of thousands of participants. The UK Biobank is one such resource that contains a substantial dataset of self-reported information on physiological, morphological, behavioral, and immunological factors. A study by Catherine Wyse and colleagues examined over 500,000 adult subject datapoints that spanned a 4-year period and identified that BMI and percent body fat are higher during the winter months and then decrease during the summer months

(Wyse et al., 2018). Data such as these reflect a large-scale meta-analysis conducted by Franz Halberg, Michel Lagoguey, and Alan Reinberg and reported in the International Journal of Chronobiology (1983). Their analyses compiled data across studies that examined seasonal peaks in several physiological processes ranging from reproduction and immunology to food intake and energy expenditure. Consistent with the UK Biobank, population trends in human body mass were identified and found to peak in September after the highest phase of carbohydrate and caloric intake (Figure 5.4). Not surprisingly, the high body mass and carbohydrate intake was found to overlap with seasonal peaks in circulating insulin concentrations. Several months after peaks in positive energy intake, there is an increase in energy expenditure represented by an increased metabolic rate late in December and early

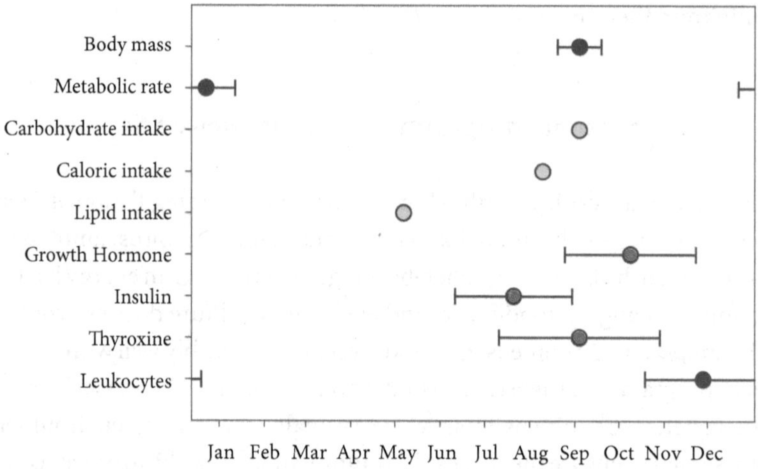

Figure 5.4 Seasonal peaks in physiological parameters associated with energy homeostasis in humans.
On average, human populations in the Northern temperate zone have the heaviest body mass in September, which coincides with higher levels of carbohydrate and caloric intake. Lipid intake and insulin precede the peak in body mass by 4 and 1 month, respectively. There is a 3-month lag in metabolic rate, suggesting that January is a month in which higher energy expenditure is engaged, which may reduce body mass. Multiple other hormonal (i.e., growth hormone, thyroxine) factors involved in energy balance have seasonal rhythms with peak concentrations in the autumn months. Error bars indicate 95% confidence intervals.
Source: Data adapted from Halberg et al., 1983.

January. The New Year resolutions to "lose weight" and "get back in shape" coincide well with the endogenous program to increase energy expenditure! The annual cycle in energy rheostasis is not an isolated physiological process, but instead is one aspect of a well-orchestrated seasonal program that also includes growth (growth hormone, thyroxine) and immune function (i.e., leukocytes).

How is seasonal or more appropriately, programmed energy rheostasis controlled in humans? And how does programmed energy rheostasis affect day-to-day variation in health and well-being? To gain the best understanding of how the central nervous system and the brain regulate energy rheostasis, we must apply August Krogh's Principle to identify the most appropriate model to study. The common biomedical mouse models have been extremely valuable for biomedical research communities due to the availability of cutting-edge genomic techniques and the ability to induce selected gene knockouts, also known as forward genetics. Unfortunately, the competitive push for rapid reproduction, development of inbred strains, and constant laboratory conditions have resulted in the loss or severe dampening of seasonal cycles in laboratory mice. Therefore, alternative animal models are required that adapt readily to laboratory conditions, maintain the naturally occurring programmed changes in seasonal physiology, show robust changes in food intake and energy expenditure to seasonal cues (i.e., photoperiod), and have neural substrates in which mediate energy balance is characterized. These conditions are met by the Siberian hamster (Stevenson et al., 2018).

During the summer hamsters maintain an average body mass of 45 grams. Over a 24-hour period, hamster food intake will be roughly 3.5–4.5 grams of regular pet-food chow. During the initial few months of autumn, the decreasing daylengths (e.g., 8hr light) will stimulate hamsters to lose approximately 30% body mass and decrease food intake 25%. These changes in body mass and food intake are due to reduced meal frequency and meal duration (Warner et al., 2010). Energy expenditure is also significantly downregulated during winter conditions. Alongside these changes in energy intake there is a 70% reduction in overall daily locomotor activity under winter conditions compared to the summer physical activity levels. Using a calorimeter that can simultaneously measure multiple metabolic parameters such as oxygen consumption and carbon dioxide production, Amy Warner working with Fran Ebling at the University of Nottingham demonstrated that hamsters in the winter short day condition display a remarkable decrease in oxygen consumption (68%) and carbon dioxide production (63%) (Warner et al., 2010).

There is also profound change in other energy output measures, such as metabolic rate.

The lower energy expenditure in Siberian hamsters under winter conditions is associated with marked decreases in the mass of multiple internal tissues, including those involved in energy balance, for example, a massive decrease of 58% in white adipose tissue and a 12.5% reduction in liver weight (Petri et al., 2014). The change in internal organ weights is not limited to those involved in energy balance, as involution is observed in the kidney (22%) and testes (94%). If humans underwent similar drastic changes in physiology and morphology, this would be equivalent to a 200lb (90kg) human dropping their body weight to 140lb (63kg) over a 2-month period. In terms of BMI, a 6ft, 200lb man would have a 27.1 BMI and be classified as "overweight." If we apply the size of the programmed changes in energy balance observed in hamsters to humans, then the equivalent decrease in humans would be a BMI score of 19.0 and verging on an underweight classification. These large-scale changes in BMI, energy expenditure, and food intake are not unique to Siberian hamsters, but are an adaptation observed in most wild rodents and large animals that inhabit high-temperate climates.

Multiple manipulations can be administered to an animal to examine how energy homeostasis is maintained. Classic approaches include those that promote "negative" energy balance or "positive" energy balance. Negative energy balance is manipulated by means that lead to a reduction in energy intake (i.e., food intake, eating) or excessive energy expenditure. Here a reduction in caloric intake or complete fasting leads to negative energy balance and results in a drive or motivation of an animal to seek food. Conversely, positive energy balance involves an over-indulgence in food intake or a significant reduction in energy expenditure. Positive energy balance is usually achieved by providing a high-fat or high-sugar meal to animals for several weeks. During the course of the new diet, animals will significantly increase body mass and fat stores in response to elevated fat/sugar intake. As we will see, manipulations of negative and positive energy balance can be used to separate energy homeostasis and programmed energy rheostasis.

Seasonally programmed energy rheostasis has some overlap with the hormonal, neuronal, and other circuits involved in energy homeostasis, the melanocortin system in particular. Yet, there are several clear genetic, cellular, and neuronal circuit differences that are specific to the rheostatic control mechanisms. As might be expected given the programmed rheostatic decreases in BMI associated with lower concentrations of leptin (Rousseau et al., 2002) and insulin (Bartness et al., 1991) during winter conditions. The photoperiodic change in circulating leptin and insulin raises the question of whether seasonal programs in body mass are driven by peripheral tissues (e.g., fat stores). If this was the case, then leptin and insulin injections to long-day animals should reduce food intake and decrease body mass. However, chronic administration of leptin to summer long-day (i.e., obese) hamsters does not reduce body mass. Similarly, if food is restricted in long-day hamsters, when the animals are subsequently "refed," there is an increase in the amount of food eaten, increased adipose tissue mass, and body mass gain. Hamsters that experienced the exact same food restriction and were then refed, yet experience leptin infusions, were also observed to increase food intake and experience body mass gain (Rousseau et al., 2002). In this case, there is a level of leptin insensitivity present, and the error detector has had a change in responsiveness that is regulated by the seasonal set point phase. These findings indicate that the peripheral tissues and derived hormones do not drive seasonal programs in body mass. Therefore, programmed rheostatic control of body mass, as with energy homeostasis, must be governed by the central nervous system, and the melanocortin system in the arcuate nucleus is a prime candidate.

Within the melanocortin system, Pomc has received the most attention for the genetic control of seasonal changes in energy balance (Helfer and Stevenson, 2020). Across many animals that exhibit programmed energy rheostasis, including bears, deer, sheep, and hamsters, Pomc is consistently observed to be expressed at higher levels when animals are in "seasonally obese" compared to "seasonally lean" states, suggesting that Pomc does not function as an anorexigenic signal at certain times of the year. This does not mean that the other neuropeptides are not responsive to energy state, as experiments that dissociate energy homeostasis from energy rheostasis reliably indicate that Npy, Agrp, and Cart monitor short-term changes in energy conditions (i.e., fasting). For example, when Siberian hamsters are housed in summer conditions with plenty of food and water, Pomc expression is higher compared to the winter short-day conditions when animals are lean and have reduced food intake (Figure 5.5). There is no day-length driven change in other neuropeptides in the arcuate nucleus involved in energy homeostasis (i.e., Cart, Npy, and Agrp). But if we remove food during the dark phase and

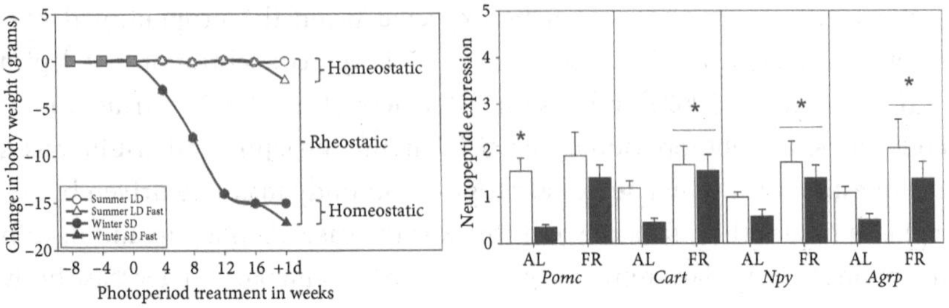

Figure 5.5 Neuronal difference between energy homeostasis and energy rheostasis.
Siberian hamsters decrease body mass in response to short winter-like day lengths. To test negative energy balance, a food restriction treatment is applied in which animals do not have access to food for one night. This experimental design allows for the separation of short-term energy homeostasis (i.e., transient food restriction) and long-term rheostasis program (i.e., seasonal rhythm) mechanisms. Both summer long-day (LD) and winter short-day (SD) animals show an approximate 2 gram decrease in body mass. The decrease in body mass is paralleled by an increase in arcuate nucleus *Npy* and *Agrp* expression. Neither Cart, *Npy*, nor *Agrp* show a programmed rheostatic change in expression. Only *Pomc* expression shows seasonal, programmed rheostasis, and not short-term homeostatic changes.
Source: Data adapted from Helfer and Stevenson, 2020.

provide an overnight fast, Pomc remains constant, yet there is a significant increase in hypothalamic Npy and Agrp expression (Figure 5.5). Findings such as these indicate that the neuronal circuit that regulates energy homeostasis remains functional across the seasons, but an additional signal must be integrated into the long-term timing in programmed energy rheostasis. The key link must be between the Pomc sensor in the arcuate nucleus and how the seasonal set point is being governed. This additional layer of programmed control is achieved via a cellular population called tanycytes that are located adjacent to the arcuate nucleus and line the third ventricle in the brain. Tanycytes are a group of glial-like cells that contact the cerebral spinal fluid, respond to circulating hormones, and send connections to adjacent neurons in the arcuate and ventromedial hypothalamic nuclei. This cell population also consists of a neural niche, capable of creating new neurons derived from stem cells. It is now well established that tanycytes are a key substrate involved in timing long-term regulated changes in physiology, in this case, energy balance.

Genetic, cellular, and neural mechanisms of energy rheostasis

There is ample evidence to implicate tanycytes in the control of energy balance (Bolborea and Dale 2013). Tanycytes contact the cerebrospinal fluid space of the third ventricle and detect and transduce information signaled by circulating amino-acids and glucose (Lazutkaite et al., 2017; Orellana et al., 2012). New findings revealed that tanycytes project directly onto Npy and Pomc cells in the arcuate nucleus to regulate their neuronal activity (Bolborea et al., 2020). In mice that are fed a high-fat diet, there is a significant increase in newly born cells and associated changes in food intake (Lee et al., 2012). Crucially, the destruction of tanycytes that connect with the arcuate nucleus increases the production of fat stores (Yoo et al., 2020). Evidence such as that described here indicates that the tanycytes can detect changes in nutrient signals circulating in the blood (i.e., detectors) and have direct connections with neuropeptides involved in the regulation of energy stores and food intake (i.e., controllers).

Professor Fran Ebling at the University of Nottingham working with Perry Barrett at the Rowett Institute of Health and Nutrition proposed that tanycytes act as a rheostat for programmed energy rheostasis (Ebling, 2015; Ebling and Barrett, 2008) (Figure 5.6). Tanycytes exhibit a remarkable level of seasonal changes; they project into the surrounding hypothalamic nuclei and down toward the pituitary gland. Using a histological labeling method that specifically identified tanycytes, research has shown that the projections retract in winter short photoperiod conditions, and a switch to long day conditions results in long radial projections that reach into the arcuate nucleus in hamsters (Kameda et al., 2003), sheep (Migaud et al., 2015), ground squirrels (Chmura et al., 2022), and rats (Helfer et al., 2016), suggesting this cell population shows an evolutionary conserved "rewiring" across seasonal timescales.

Regulation of the predictable change in tanycyte morphology is also an evolutionary conserved mechanism, and starts with an organ in the neck. The thyroid gland has long been known to influence energy balance by regulating metabolic rate and plays a critical role in growth and development. This is achieved by the secretion of two hormones: thyroxine (T4) and triiodothyronine (T3). The secretion of both hormones is regulated by a peptide that is produced in the pituitary gland, thyrotropin-stimulating hormone (*Tshβ*). But what does the pituitary-thyroid gland connection have to do with

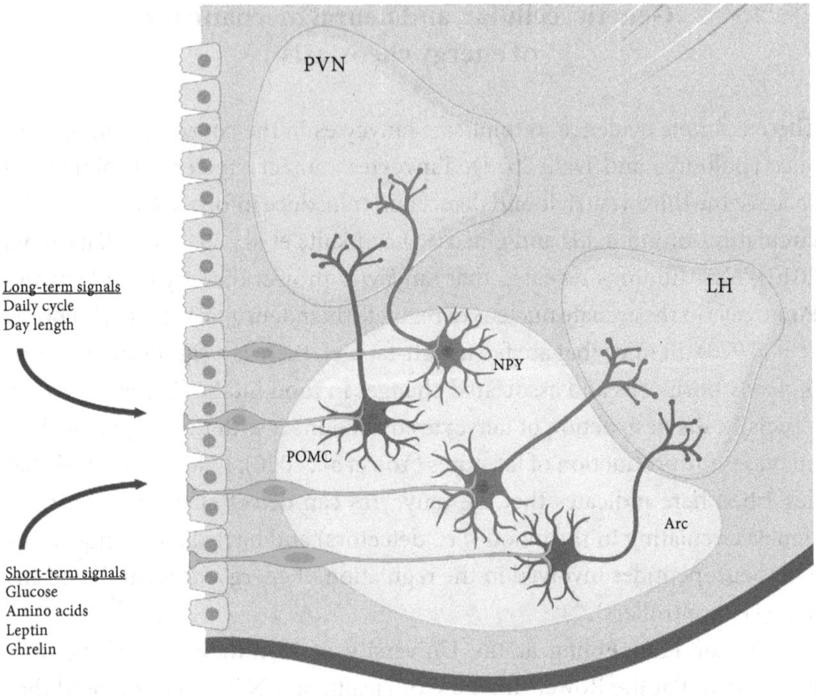

Figure 5.6 Tanycytes orchestrate programmed energy rheostasis. Tanycytes along the ependymal layer in the third ventricle of the brain coordinate the long-term regulated changes in body mass. Tanycyte cells, denoted in green, detect long- and short-term signals via the cerebrospinal fluid in the ventricle. Short-term signals derived from food intake are detected by tanycytes and these cells project to first order neuropeptides to provide an additional pathway to signal energy homeostasis. Long-term signals derived from daily and seasonal cues are integrated by tanycytes and entrain an endogenous circannual interval timer to control programmed energy rheostasis.

tanycytes located deep in the brain? Two special types of molecules called enzymes are involved in changing the structure of thyroid hormones into "biologically active" and "biologically inactive" forms. These enzymes, *Dio2* and *Dio3*, are coded by the genes deiodinase Type-2 and deiodinase Type-3, respectively, which are expressed in cells located in many peripheral tissues as well as the brain. Dio2 and Dio3 function to control the amount of the biologically active thyroid hormone (T3) available in each tissue/organ. The enzyme Dio2 converts T4 into the more potent hormone, T3. Conversely, the enzyme Dio3 converts both T4 and T3 into biologically inactive hormones. These enzymes are critical for producing biologically active hormone required to cause neuro-morphological changes in tanycytes and perhaps counterintuitively, for stopping or preventing hormone action in sensitive brain regions. Tanycytes express both Dio2 and Dio3 enzymes, and this expression shows an endogenous long-term cycle that is sensitive to changes in daylength. Within several seasonal animals, tanycytes show a remarkable seasonal change in the expression levels of these enzymes and the production of thyroid hormone (i.e., T3) locally in the brain (Yoshimura, 2013; Dardente et al., 2014).

In summer long day (LD) conditions, the local production of hypothalamic thyroid hormone by tanycytes increases food intake and elevates body weight. In winter short day (SD) conditions, there is a cessation of thyroid hormone production in tanycytes and a significant reduction in food intake and body weight. A wealth of data collected from Siberian hamsters has shown that summer LD conditions are associated with higher Dio2 expression in the tanycytes. Once hamsters are moved to the winter SD-like conditions the expression of Dio3 significantly increases in the tanycytes and both T4 and T3 are converted into the inactive hormones—reverse triiodothyronine (rT3) and diiodothyronine T2, respectively. The reduction in hypothalamic thyroid hormones T4 and T3 results in body mass loss, food intake, and energy expenditure. Perry Barrett and colleagues at the Rowett Institute conducted a brilliant study to confirm the sufficiency of T3 to regulate Siberian hamster body mass (Barrett et al., 2007). The logic of the study was: if thyroid hormones maintain the higher body mass, then hamsters placed in the SD condition and receiving thyroid hormone injections should still have high body mass despite exposure to the short winter days. Indeed, this is the case, hamsters that were transferred to short days displayed the typical loss in body mass, yet those with T3 micro-implants in the hypothalamus failed to reduce body mass. These data show that the tanycytes' production of thyroid hormone is a key step for the long-term programmed energy rheostasis.

Studies that examine how day length regulates Siberian hamster body mass usually focus on relatively short-term (e.g., 6–12 week) changes in response to short days. What happens when hamsters are exposed to short days for longer than 12 weeks? After around 20 weeks of short days, hamsters will spontaneously change and revert to the LD phenotype even though they are still exposed to short days! Hamsters in this state have an endogenous switch in the central nervous system and become physiologically unresponsive to the SD photoperiod. The endogenous, *circannual* induced change in physiology results in hamsters are indistinguishable from those animals kept in LD conditions. What happens in the tanycytes of these hamsters? Well as might be expected, there is a decrease in Dio3 expression and a resumption in the ability of Dio2 to create T3 that acts in the hypothalamus (Stevenson and Prendergast, 2013). In hamsters that are kept in a semi-natural environment, a constellation of genomic changes in tanycytes, including *Dio2* and *Dio3*, cycle and function to orchestrate the seasonal timing of programmed rheostatic changes in body mass (Petri et al., 2016). The spontaneous reversion from the SD winter condition to the LD summer condition, even though the hamsters are still experiencing short days, is an example of an annual interval timer. In this case the increase in hamster body mass and food intake in SD conditions serve to anticipate long days and helps the animals prepare for summer.

The European hamster (*Cricetus cricetus*) is another powerful model for examining the endogenous circannual timing of energy rheostasis due to the ability to identify robust oscillations in its body mass across a simulated annual period. Male hamsters that are kept in a constant day length show strong cycles in *Dio2* expression when comparing the subjective summer versus the subjective winter, with higher expression coinciding with larger body mass (Saenz de Miera et al., 2014). These findings provide tantalizing evidence that thyroid hormone concentrations oscillate across the year to trigger seasonal, phase specific changes in tanycyte function. It is likely that other factors are also integrated by tanycytes to coordinate the long-term changes in energy rheostasis (e.g., glucose, amino acids).

Until now we have focused predominantly on hamsters due to their robust responses to day length changes. But is there an animal model that can help us better understand the molecular basis of circannual timing of body mass? Here the golden-mantled ground squirrel (*Callospermophilus lateralis*) is another suitable organism according to the August Krogh's Principle. The circannual changes in body mass, reproduction, and hibernation of golden-mantled squirrels have been studied for over 40 years (Kenagy, 1980). The circannual change in squirrel body mass is mirrored by food intake (Zucker and Boshes, 1982). Remarkably, squirrels with SCN lesions to disrupt light and day length cues had no effect on these circannual rhythms in food intake and body mass (Dark et al., 1985). These data indicate that although the SCN integrates daily light:dark cycles and the measurement of photoperiodic cues can fine tune the timing of body mass cycle, the endogenous circannual timer for programmed energy rheostasis does not require SCN-dependent day length information. Arctic ground squirrels show a programmed rheostatic change in response to short days like the one described for Siberian hamsters. Recent work by Helen Chmura working with Cory Williams at Colorado State University showed that these animals also exhibit remarkable tanycyte rewiring that serves to time annual changes in energy balance, hibernation, and arousal (Chmura et al., 2022). The current evidence indicates that tanycytes are the predominant cell population that governs programmed energy rheostasis in body mass, as well as other oscillations in seasonal physiology (e.g., reproduction) (Lewis and Ebling, 2018).

Framework for programmed energy rheostasis

How do these genetic, cellular, and hypothalamic connections that govern seasonal cycles in energy balance function within a programmed rheostatic mechanism? Mrosovsky proposed that rheostatic processes, particularly those involved in the control of hibernation, used a high- and a low-level detector. Both detectors, when signaled by deviations in the regulated variable, converged on an effector. The central difference between rheostatic and homeostatic processes was the absence of a set point. With the advances in genetic and cellular discoveries over the past few decades, a rheostatic mechanism can be identified that functions in a manner akin to homeostatic systems. The inclusion of a set point that oscillates between "high" and "low" level detectors removes the need for two separate entities. A revised model for programmed energy rheostasis is outlined in Figure 5.7. The central premise of the rheostatic model is that the underlying stimulus-response homeostatic processes are maintained, and the long-term regulated changes occur at a *secondary* higher order level. The inclusion of the homeostatic model above is justified as all evidence indicates that the same sensory, error detectors, controllers, and effectors are necessary for rheostatic processes. In the case of seasonal rhythms in body mass as a regulated variable, the primary effectors are also peripheral tissues that signal energy sources (e.g., stomach/adipose/pancreas). The dorsal vagal complex is the controller for the effectors and functions to maintain

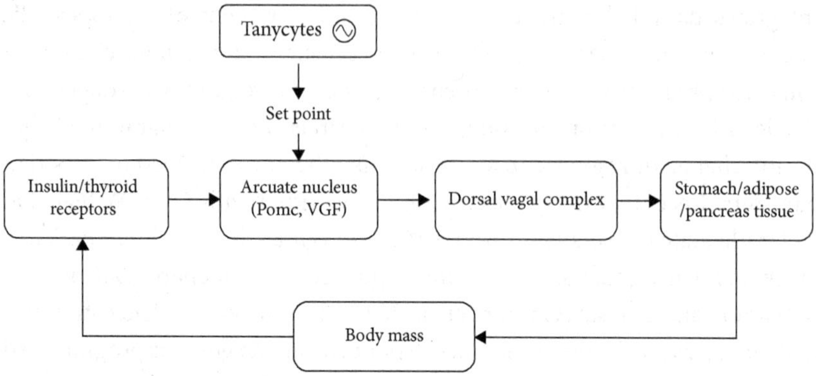

Figure 5.7 Schematic representation of programmed energy rheostasis. Annual rhythms in body mass are evident across the animal kingdom. Most animals show seasonal oscillations in both energy intake (i.e., food intake) and energy expenditure (e.g., physical activity). Across mammal and bird species, *Pomc* cells in the arcuate nucleus exhibit robust rhythmicity in expression with seasonal peaks and troughs that are driven by day length cues. Tanycyte cells that line the third ventricle in the hypothalamus integrate multiple environmental and internal cues and display seasonal changes in morphology. The extension and retraction of connections with neurons in the hypothalamus and in the median eminence allows tanycytes to convey time-of-year information and thus regulate the set point that is established by neurons involved in energy homeostasis (i.e., Pomc). Evidence suggests that tanycytes have an endogenous circannual timing mechanism indicated by the oscillation icon.

rheostatic control over food intake and energy expenditure. The primary difference between short-term homeostatic and long-term rheostatic is the error detector requires far fewer neuropeptides, such as Pomc, and not the full set of first order (and associated) central and peripheral orexigenic and anorexigenic peptides; Pomc expression is insensitive to short-term energy signals (i.e., fasting), yet reliably codes for long-term changes in the regulated variable, body mass. Like the programmed circadian rheostatic model described in Chapter 3, specific nuclei and cellular substrates provide direct (and potentially indirect) input to error detectors (i.e., Pomc). Here tanycytes provide circannual timing and code for the endogenous annual set point to transition between "high" and "low" detection limits. Tanycytes directly connect with Pomc expressing cells in the arcuate nucleus to drive the seasonal programmed energy rheostasis.

Summary

In this chapter, the long-term control of seasonal cycles in energy balance was presented and focused on the genetic, cellular, and neuronal circuit differences between homeostatic and rheostatic processes. For seasonal rhythms in physiology, the tanycytes cell population along the third ventricle plays a pivotal role in the control of long-term oscillations in multiple physiological systems. Here, the focus was on programmed energy rheostasis, but tanycytes are involved in annual set points for most physiological systems governed by the hypothalamus. Tanycytes are well positioned to integrate both environmental (e.g., day length) and internal (e.g., nutrient) signals and orchestrate the plethora of neuroendocrine plasticity involved in seasonal rhythmicity. The extensive retraction and projection of tanycyte connections with adjacent neurons (i.e., Pomc) and median eminence show that those programmed changes in these glial-to-glial and glial-to-neuron connections are a key difference in the circuit involved in error detection and controller function. Given that the seasonal variation in physiology can occur independently of external cues, evidenced by the rhythms in ground squirrel body mass (Kenagy, 1980) and migration in birds (Gwinner and Dittami, 1990), the programmed changes in tanycyte morphology and function are likely driven by an uncharacterized circannual mechanism like the circadian clock. Undoubtedly, the genetic and/or cellular property of circannual timers will be resolved in the not-so-distant future, perhaps by an inquisitive reader.

We have now covered three clear examples of long-term programmed changes in physiological systems: (1) circadian clocks, (2) female reproductive cycles, and (3) seasonal cycles in energy stability. For each example of programmed rheostasis in Chapters 3–5 the genetic, cellular, and neuronal network involved in the long-term regulated changes in physiology were outlined and shown to be dissociated from the short-term homeostatic systems. The fundamental basis for programmed rheostatic changes is the endogenous pacemakers that maintain a clear periodicity in the long-term rhythms. One underlying feature is that homeostatic systems are maintained despite the marked phase shifts in the timing of rheostatic programs. In Chapters 2 and 3, the impact of disrupted rhythmicity was shown to have a clear detrimental impact on rheostatic processes, even though homeostatic responses remained intact. One challenge in the field of seasonal physiology is to examine the consequences of disrupted programmed changes on long-term fitness. The experiments required are immensely time consuming, require funding commitments well beyond the typical grant life cycle (of usually 3–5 years), and expire long after the career of most researchers. Now that we have established the properties of programmed rheostasis, let us turn our attention to a very different type of regulated change in physiology variables—reactive rheostasis.

Note

1. Sir Frederick Banting's life and research career were wracked with struggles. He failed his first year in the General Arts program at Victoria College, University of Toronto. In 1914 he was refused enlistment in the army due to poor eyesight. After the war he was unable to gain employment at the Hospital for Sick Children in Toronto. He then moved to London, Ontario, and needed to lecture at the University of Western Ontario to supplement his income as his medical practice was not particularly successful. Finally, he struggled to duplicate his first successes in isolating insulin!

6
Stability during recovery

I'll be here when it all gets weird; If I ever leave this world alive.
Flogging Molly (1997–present)

In 2020, the world became inundated with information on viral infection and transmission due to the highly contagious acute respiratory illness caused by a coronavirus (COVID-19). After becoming infected with the virus there is an incubation period of on average 5 days, and clinical symptoms develop in most people (i.e., 98%) within 11 days. The most common symptoms of people infected with COVID-19 was fever (approximately 88%), followed by a dry cough (68%), tiredness (38%), shortness of breath (19%), muscle aches (15%), sore throat (14%), and headache (14%) (Tay et al., 2020). Some of these symptoms overlap with seasonal flu (i.e., influenza) including fever, dry cough, and headache, but COVID-19 infection also included shortness of breath, and influenza is generally associated with diarrhoea. Once the symptoms of COVID-19 were observed, fever, sore throat, and aches generally lasted around 5–6 days, whereas a cough was observed for approximately 11 days. The total period of contagiousness of an individual was roughly 2 weeks. In severe cases (2% of patients), hospital admission was necessary, and acute respiratory distress syndrome manifested between 8 and 11 days. The common symptom observed across COVID-19 and other similar viral infections (e.g., influenza) is fever, a normal and evolutionary ancient physiological response that functions to help the immune system defend against an attack from an infectious organism. Commonly observed during an infection is a well-orchestrated, regulated change in the onset, duration, and termination of the fever response.

In the previous three chapters, the focus revolved around timed, rhythmic changes in physiological variables that cannot be accounted for using the well-described short-term homeostatic mechanisms. Reactive rheostatic responses, however, are not only involved in regular patterned activity but

On Rheostasis. Tyler John Stevenson, Oxford University Press. © Oxford University Press 2024.
DOI: 10.1093/oso/9780197665572.003.0006

can also be involved in the regulation of systems that exhibit non-rhythmic patterns and instead show predictable regulated physiological changes that span a single period or *bout*. In these situations, as outlined in Chapter 2, reactive rheostatic mechanisms are best suited to explaining regulated changes in physiological variables despite continued first order homeostatic mechanisms. Here, the precise timing in the length of a fever is presented as a clear manifestation of a regulated series of physiological responses. The induction and maintenance of fever during an immune response is an excellent example of a reactive rheostatic change in a physiological system: core body temperature.

It's getting hot in here: fever is an evolutionary ancient response

Mammals and birds are homeothermic, meaning that they attempt to keep their body temperature relatively stable around a set point. Homeothermy has evolved in the face of abundant energy stores which facilitated self-regulation of internal body temperature to optimize biochemical reactions. In most mammals the set point for body temperature is significantly higher than the external temperature at approximately 37°C. Fever or pyrexia, from the Greek word for pyros or fire (Kluger, 1979), is typically defined as a sustained, regulated elevation in core body temperature of roughly 1.5°C and has been a primary tool used in clinical settings for monitoring disease progression for over 2000 years (Coxe, 1846). Most infectious organisms have also evolved to have optimal growth rates at normal mammalian body temperatures (~37°C). Where a fever is present and body temperature increases to 39°C–40°C, the environment becomes inhospitable for infectious pathogens, such as COVID-19, which slows further multiplication and causes destruction of the infectious agent. In the event an infectious organism enters the body and is detected, there is a rapid immune response that is often accompanied by a change in the set point for body temperature (i.e., homeostasis) to a higher regulated temperature (i.e., rheostasis). In most mammals and birds, a fever response can be detected 30 minutes after infection and can last up to 9–14 hours (Figure 6.1).

Laboratory rabbits (*Oryctolagus cuniculus*) were historically used to study fever responses due to the rapid and robust change in body temperature after an injection of a pathogen that mimicked infection (Kimura et al., 1994).

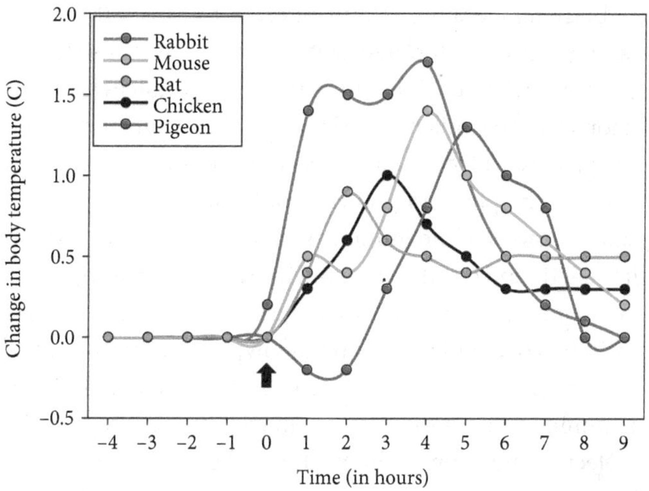

Figure 6.1 Comparative perspective on fever in birds and mammals. Day to day changes in core body temperature are extremely small, no larger than the naturally occurring daily rhythms. An injection of a pathogen at the time indicated by the black arrow induces an increase in core body temperature. A sustained 1–1.5°C increase in core body temperature is characteristic of a fever state and is maintained for several hours (e.g., 5–9). Fever helps homeotherms (i.e., birds and mammals) to fight infection and has significant adaptive value.

Within minutes there is an increase in core body temperature that lasts 5–6 days. Now, mouse models are widely used due to the ability to link specific genes with any physiological response. In rodents, there is a remarkable similarity in the fever bout length and increased core body temperature with a slight delay compared to rabbits, but a prolonged elevation for over 7–8 days (Kozak et al., 1994; Prendergast et al., 2015). In other homeotherms, including chickens (D'Alecy and Kluger, 1975) and pigeons (Nomoto, 1996), core body temperature is raised for a near identical length of time.

Fever is an evolutionary conserved physiological response that is widely observed across the animal kingdom, including ectotherms (i.e., cold blooded lizards). Ectothermic animals do not have physiological control over the fever response but will move toward a heat source to elevate body temperature, a process called behavioral thermoregulation (Kluger, 1978). Even though these "cold-blooded" animals are incapable of maintaining a core body temperature set point, they use the environmental temperature to regulate body temperature, a trait referred to as ectothermy (from the Greek meaning outside [ektos] and heat [thermos]). Both fish (Reynolds and Casterlin, 1979) and lizards (Vaughn et al., 1974) exhibit behavioral thermoregulation for a period similar to the length of the mammalian fever.

In this chapter, we will focus on the reactive rheostatic control of core body temperature and the physiological control of the fever response. First, the anatomy and divisions of the immune system will be described followed by the specific genetic and cellular components that comprise aspects of the immune response pertinent to fever. Then the hormonal mechanisms that control fever will be covered followed by the hypothalamic control of effector tissues via the peripheral nervous system. The aim is to highlight that reactive innate immune rheostasis involves an evolutionary conserved series of discrete molecular, hormonal, and anatomical structures that are activated over a relatively long term (e.g., days to weeks) to induce a highly regulated change in core body temperature.

The anatomy of defensive systems in mammals

The first line of defense against infectious organisms is typically physical barriers, including skin or exoskeleton structures such as those used by insects. Other natural barriers that prevent organisms from invading our bodies include the mucous membranes that line the cavities in the body: tears on our eyes, earwax, and the highly acidic environment in our stomachs. In the event an organism can evade these structures and environments, two types of immune responses are recruited to aid the body's defense. The first is called the *innate* immune response, because it includes genetically inherited receptors that recognize pathogens such as bacteria, viruses, fungus, and other foreign organisms that the body does not recognize (e.g., parasites). The second type is *adaptive* immunity, which does not involve genomic inheritance but instead consists of a physiological network that "learns" to identify foreign molecules that are present on the surface of a pathogen. For adaptive immunity, the body creates antibodies that specifically recognize or "remember" a specific pathogen (also known as an antigen) and recruits immune cells to help remove and destroy the invasive organism (Dempsey et al., 2003), such as the virus which caused the coronavirus pandemic in 2019, SARS-CoV-2 or severe acute respiratory syndrome coronavirus 2.

The fever response has bi-directional interactions with multiple peripheral tissues that are involved in innate and adaptive immunity (Figure 6.2). The lymphatic system is a series of organs and fluids that interact with both the circulatory and immune systems. The lymphatic system functions by moving an aqueous filtrate of blood called lymph through the lymphatic and lymph nodes before returning the fluid to the circulatory system. Lymph is a colorless watery solution (named after the Roman deity of fresh water,

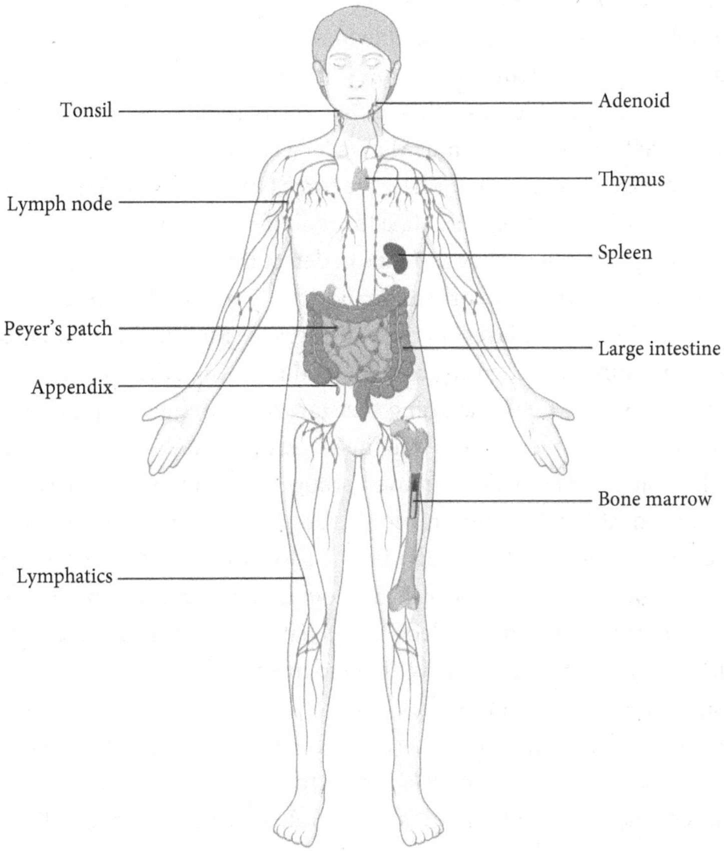

Figure 6.2 Anatomy of the immune system.
Immune cells are derived from stems cells located in bone marrow that differentiate into B-cells or are released into the circulatory system and develop into T-cells in the thymus. Peripheral lymphoid tissues include the adenoid, tonsil, lymph nodes, spleen, Peyer's patch in the small intestine, and appendix. The lymphatics carry active dendritic and macrophage cells to the lymph nodes to remove pathogens.

Lympha) that contains immune cells that attack and destroy infectious bacterial and viral pathogens. Within the lymphatic system there are two classes of organs called the primary and secondary lymphoid organs. The primary lymphoid organs include the bone marrow and thymus. Inside the bone, or the marrow, is a unique environment called a *niche*, which contains a group of undifferentiated cells called hematopoietic stem cells that are capable of maturing into any cell found in the blood and all immune cells (Wilson and Trumpp, 2006). While these stem cells hold significant promise for the treatment of acute and chronic diseases, due to their ability to transform into any blood cell, so far, no one has been able to control and grow these cells in cell culture. The bone marrow and thymus are responsible for the production of lymphocytes, classified as either B- or T-cells depending on the tissue in which the cell "matures." Thymus-derived T-cells are critical for both innate and adaptive immune systems. There are three types of B-cells—memory, cytotoxic, and helper—which circulate through the body and monitor all cells for any changes in cell surface antigens to detect infectious pathogens.

The secondary lymphoid organs include the lymph nodes, spleen, tonsils, and mucous membrane layers in the body organs (Ruddle and Akirav 2009); in them, immune cells identify, combat, and remove pathogens. The spleen has a wide range of roles, including the storage of immune cells, and in response to infection releases cells into the blood or produces a type of immune cell called phagocytes, which remove pathogens from the bloodstream. The tonsils are located bilaterally in the throat and contain a high number of leukocytes (i.e., white blood cells) that act to stop pathogens entering the body through the mouth or nose. Lymphoid tissue embedded in the mucosa layer of the small intestine, called Peyer's patches, form follicles consisting of different types of immune cells; these identify and defend the gut against pathogens and produce antibodies. Infections that are identified in any primary or secondary lymphoid tissue can induce a fever response, for example, when the bacteria *Salmonella typhi* infects the intestine tract a severe pyretic (i.e., fever) response results, known as typhoid fever (Everest et al., 2001).

Cellular basis of pathogen recognition and attack

Multiple immune tissues can be involved in the initiation of a fever response, but the primary means of detecting infected cells and mounting an immune response resides in a specific class of cells that circulate throughout the body.

The lymphoid lineage of bone marrow derived cells includes natural killer cells, T-lymphocytes, and B-lymphocytes. Natural killer cells are the attack arm of the immune system, as they recognize molecules on the surface of infected cells and destroy them by releasing molecules that induce lysis of the pathogens (Björkström et al., 2016). Natural killer cells are activated by cytokines and interferons secreted by infected cells, macrophages, or dendritic cells. These cells are one of the first lines of action to prevent the spread of infections until reinforcements (from the adaptive immune system) arrive. Natural killer cells also express receptors on their surface that prevent them from destroying cells produced by the host, known as "self-molecules." T- and B-lymphocytes have different roles during an immune response and are broadly characterized as killers or helpers, and both are producers of antibodies, respectively (Ratajczak et al., 2018). The myeloid lineage of bone marrow, on the other hand, gives rise to the erythrocytes (i.e., red blood cells), mast cells (e.g., inflammation associated with allergies), and myeloblasts. Myeloblast cells are further differentiated into basophils, neutrophils, and eosinophils. The 1908 Nobel Prize in Physiology or Medicine was awarded to Paul Ehrlich and Ilya Mechnikov for their discoveries characterizing basophils, neutrophils, and eosinophils and elucidating their function in immunity (Valent et al., 2016). Basophils primarily induce an inflammatory response that is associated with allergic reactions. Neutrophils are more abundant in the circulatory systems and are involved in general inflammatory responses. Eosinophils are produced in low numbers and attack infectious organisms such as parasites. The primary group of immune cells essential for the fever response and common across the animal kingdom are monocytes and will be the focus for the rest of this chapter.

Monocytes are granulocytic leukocytes that can differentiate into macrophages and dendritic cells. Macrophages are critical for the detection, phagocytosis (i.e., "eating" of a pathogen), and destruction of infectious organisms such as bacteria by antibody-dependent cell-mediated cytotoxicity. In addition to their role in innate immune responses, macrophages initiate a local inflammatory response and induce fever responses via released immune "hormones" (i.e., cytokines). Dendritic cells are also involved in the innate immune response, as well as the adaptive immune response; they were discovered by Ralph Steinman and Zanvil Cohn in 1973 using a combination of cell culture assays and animal models that focused on immune cells derived from mouse spleens. Ralph Steinman, a Canadian immunologist working at the Rockefeller University in New York, established that dendritic cells could be dissociated from macrophages as they express a key receptor known as major histocompatibility complex II and function during the innate immune response to release interferons and cytokines, but also activate adaptive immune responses by "presenting" pathogens to T-cells and establish immune memory. His work was recognized with the Nobel Prize in Physiology or Medicine in 2011.[1]

Innate mechanisms to recognize self and non-self

But, how do macrophages and dendritic cells "recognize" an infectious pathogen? This is especially important as there is considerable variation in the morphology of bacterial and viral structures. Arguably, one of the most important immunological discoveries of the twentieth century was that macrophages and dendritic cells recognize infectious organisms via Toll-like receptors, which then trigger the immune response. Perhaps it is not surprising that the Nobel Committee opted to award the Nobel Prize in Physiology or Medicine to Jules Hoffmann and Bruce Beutler for this discovery. Toll-like receptors are widely observed in the animal kingdom, but even more remarkable, there is a homologous or highly similar receptor in plants that also confers disease resistance. Comparison of sequences of Toll-like receptors across the animal and plant kingdom suggests that receptors that provide defense to invasive organisms evolved over 2 billion years ago (Kimbrell and Beutler, 2001).

Toll-like receptors are present on the surface of cells and act as a sensor for the innate immune system. Their activation results in the production of other immune genes in the cell, such as inflammatory cytokines and interferon transcripts (Janeway and Medzhitov, 2002). The unique ability of Toll-like receptors to recognize "non-self" from "self" molecules allows the macrophages and dendrites to determine and target pathogens, but not the organism's own cells. The Toll-like receptors on the outside of the cells are complemented by other immune sensors of the innate immune system that are located inside cells, including genetically encoded molecules such as retinoid acid-inducible gene receptors and oligomerization domain-like receptors. These intracellular receptors recognize viruses and other infectious pathogens and rapidly activate immune genes to induce a complex response that destroys and removes the pathogen (Rehwinkel and Gack, 2020).

If cells are infected with bacteria or viruses, several conserved immune genes are induced and function to not only support the destruction of the pathogen but also provide a hormone signal to adjacent cells and support adjacent and nearby cells to mount a local immune response. Infection leads to acute inflammation, a process that is relatively short in duration and is characterized by the weakening of the blood vessel walls so that immune cells (i.e., white blood cells), plasma proteins, and fluid are increased locally at the site of infection. This localized acute inflammation serves to immediately increase the temperature around the infection to prevent the pathogenic organisms spreading to other body parts. One of the earliest responses of infected cells is to secrete interferons (Figure 6.3). Toll-like receptors that are activated by viral or bacterial pathogens activate intracellular signaling pathways that initiate the synthesis of Type I interferons, cytokines, and

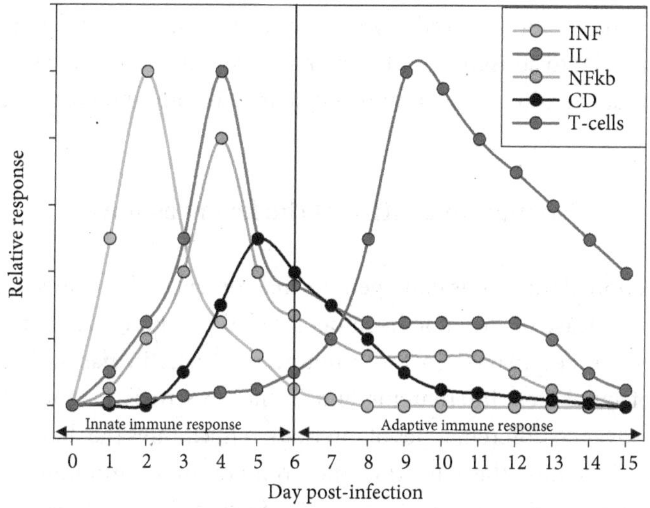

Figure 6.3 Hormonal basis of reactive rheostasis for the immune response. Innate immune responses are activated in response to the detection of pathogens (e.g., bacteria) and induce a long-term regulated change in several hormones. Early in the reactive rheostatic response, interferon (INF) molecules are produced and secreted into the circulatory system. Cytokines such as interleukins (IL) are recruited along with necrosis factor kappa beta (NFkb). Cluster of differentiation (CD) molecules are then increased and peak 5 days post-infection. Infections that trigger adaptive immune response involve high levels of T-cells (and B-cell) production. The innate immune response typically lasts 5–7 days.

chemokines (Borden et al., 2007). Type I interferon genes act to protect the cell against acute viral infections and are also involved in some bacterial infections (MacMicking, 2012). Interferon genes also stimulate the expression of nearly 2000 genes and trigger an anti-pathogen response that attacks the infectious organism. Induced interferons have evolved recently and are present in vertebrates (MacMicking, 2012). Interferons provide cell-autonomous host defense mechanisms and are the initial line of defense after an infectious organism penetrates inside the body. When infections spread beyond a local region, systemic detection and activation of the innate immune response results, and the fever response is quickly activated to prevent further organismal-wide infection. The macrophage and dendritic cells are critical in signaling that the infection has spread beyond a localized region and the secretion of "immune hormones" that initiate a fever response. The activation of the innate immune response involves a constellation of specialized cell populations and hormones. A full description of the steps of the innate immune response is well beyond the scope of this chapter, and readers are encouraged to consult any edition of *Janeway's immunobiology* for an advanced understanding of the immune system (Murphy and Weaver, 2016).

Hormonal control of the fever response

The detection of infectious pathogens by macrophage cells results in the synthesis and release of immune hormones called cytokines. As noted above, cytokines are key players in the immune response, in particular specific classes of cytokines: interleukins and Tumor Necrosis Factor-α (TNFα). There are over 40 different types of interleukin hormones that have a range of functions, including the activation and proliferation of immune cells, stimulating the migration of cells to sites of infection. Interleukins are quickly produced by macrophages and released into the circulatory system to induce a local and systemic pro-inflammatory response. Interleukin-1 was the first described in 1977 and consists of two proteins, an alpha and beta form (Dinarello et al., 1977). Interleukin-1 is released by macrophages into the circulatory system approximately 2–3 hours after an infection and peaks 4 days post-infection. One of the main targets of interleukin-1 are the error detectors of the reactive rheostatic fever response located in the brain. TNFα was first identified in the 1970s, found in large concentrations in the blood of mice infected with a common bacteria, lipopolysaccharide, and observed to cause hemorrhagic necrosis of cells (Carswell et al., 1975). TNFα is released 2–3 hours after detection of a pathogen and acts on adjacent blood vessels to produce adhesion molecules that bind to white blood cells (i.e., leukocytes). The leukocytes and neutrophils then move out of the blood vessels at the site of infection and facilitate in the destruction and removal of infected cells. TNFα induces leaky blood vessels to clot, which prevents further fluid loss but also results in the local inflammatory response associated with heat, pain, and swelling 3–4 days post-infection (Figure 6.3) (Kalliolias and Ivashkiv, 2016). The hormonal control of fever is generally consistent across innate and adaptive immune responses and last approximately 3–7 days. In the case that an adaptive immune response is engaged, the system generates "memory" antibodies, in part from the action of a cluster of differentiation molecules and the increased production of T-cells by 9 days post-infection (Figure 6.3). If we return to COVID-19 infections, cluster differentiation 4+ (CD4+) expressing cells function to remove the virus from the body, whereas CD8+ expressing cells provide protection from viral infection (Sette and Crotty, 2021). Indeed, hundreds of different general defensive molecules as well as antigen-specific memory cells are recruited to mount an effective response. For the purpose of the hormonal control of the reactive rheostatic fever response, both circulating interleukin-1 and interleukin-6 produced from macrophages and, to some extent, dendritic cells are the main signal that triggers the long-term regulated change in core body temperature physiology.

Interleukin-1β and interleukin-6 are classed as endogenous pyrogens because they are produced by immune cells located within the body and function to induce a fever through triggering a reactive rheostatic response. Fever can also be induced by exogenous pyrogens, which are substances released by externally derived infectious organisms (i.e., bacteria) that also cause interleukin release. Both endogenous and exogenous pyrogens act on a small cell population in the brain, specifically in the hypothalamus, to stimulate the production of a powerful hormone, prostaglandin E2 (PGE2), the primary cause of inflammation and fever in mammals (Park et al., 2006). Increased synthesis of PGE2 in the hypothalamus signals the induction of fever by increasing heat production by effector tissues (e.g., brown fat) and preventing heat loss (e.g., increased peripheral vasoconstriction). The clinical management of fever and peripheral inflammation has for more than 100 years been achieved using the salicylic acid derivative aspirin, as this drug acts to prevent the induction of prostaglandin (Vane and Botting, 2003).

Neuroanatomical structures for fever-induced control of core body temperature

The primary cell group responsible for detecting circulating interleukins within the hypothalamus are the endothelial cells that surround key brain regions known as the circumventricular organs, where communication between the circulatory system and the cerebrospinal fluid is possible, and which provide nutrients to the brain. Critically these endothelial cells integrate the hormonal signals of infection in the peripheral blood system and signal in the hypothalamus to increase core body temperature (Figure 6.4). This is achieved by macrophage-derived interleukin-1β stimulating endothelial cells that surround the hypothalamus to transcribe a gene called cyclooxygenase 2 (*Cox2*) (Cao et a., 2001). The enzyme COX2 changes the structure of other molecules (e.g., arachidonic acid) inside the endothelial cells into prostaglandin E2 (PGE2) (Seibert et al., 1994; Masferrer et al., 1994), which is released and travels to adjacent neurons located in the preoptic area where it binds to the EP3 receptor. These are located on temperature sensitive neurons, which act as a controller to regulate the increased core body temperature (Lazarus et al., 2007). PGE3 binds to EP3 receptors on warm-sensitive neurons (Ranels and Griffin, 2003), which co-express GABA, an inhibitory neuron that acts to engage the peripheral nervous system. The binding of EP3 expressing neurons also induces the sympathetic nervous system to release noradrenaline, which increases core body temperature by thermogenesis in brown adipose tissue and induces peripheral vasoconstriction to prevent heat loss (Blomqvist and Engblom, 2018). This response pathway can also be supplemented by the actions of interleukin-6, which stimulates astrocytes to produce nuclear factor kappa light chain enhancer of activated B (*NFκβ*) which then also acts on the endothelial cells to stimulate synthesis of *Cox2* and its associated downstream events.

Now that we have established the temporal sequence of physiological stages that govern the regulated change in core body temperature during the fever response, let's turn to how the reactive rheostatic and homeostatic responses can be dissociated. One test that has been used to effectively dissociate the two physiological systems involves an experiment that uses humans, a large bath, a pathogen (i.e., Propidon), and a glove (Cabanac and Massonnet, 1974). There were two experimental conditions. In one, which served as the control, human subjects were immersed up to their chin in a large bath held at either 30°C or 40°C. Participants were then required to keep

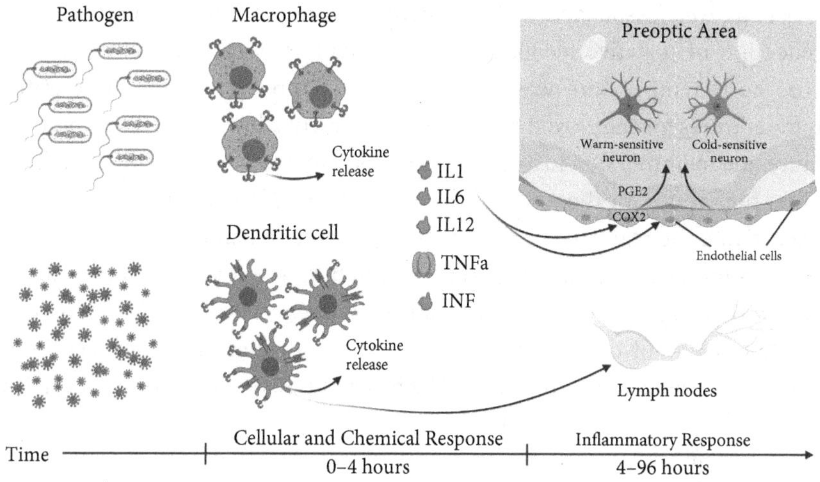

Figure 6.4 Cytokine and neuronal substrates for the control of fever. Infections caused by invasive bacteria and viruses are identified by white blood cells, including macrophages and immature dendritic cells. Bacteria are identified by toll-like receptors on the surface of macrophages, which stimulate the synthesis of cytokines interleukin-1 (IL1), -6 (IL6), and -12 (IL12) as well as tumor necrosis factor alpha (TNFa). Immature dendritic cells identify both bacterial and viral pathogens; their activation results in the synthesis of interleukins and interferons (INF; in the case of viral infection) and production of major histocompatibility complex genes. Cytokines travel through the circulatory system and stimulate endothelial cells adjacent to the brain region close to the preoptic area in the hypothalamus to increase cyclooxygenase 2 (COX2) expression. COX2 activity increases the local concentrations of prostaglandin E2 (*PGE2*), which acts on temperature sensitive neurons in the preoptic area to initiate fever responses. Mature dendritic cells take viruses and bacteria to lymph nodes and engage other white blood cells (e.g., T-cells) to facilitate infectious pathogen removal.

their left hand above the water, and a glove was placed over the hand. Water was moved into the gloved hand, and the temperature was selected by the participant. The core body temperature was monitored from the esophagus, and skin temperature was recorded from the hand in the glove. The results of the two bath temperatures indicated that the participant could change the water temperature in the gloved hand as the body attempted to "self- regulate" the core body temperature through peripheral changes in blood flow. As would be predicted based on homeostatic control, a core body temperature was negatively associated with skin temperature (i.e., gloved hand), a clear reflection of negative feedback control (Figure 6.5). Specifically, a higher core body temperature was associated with a preference for lower water temperatures for the gloved hand. And conversely, a lower core body temperature due to immersion in a colder bath was associated with a higher temperature preference for the water in the gloved hand. Now, what happens when we activate the reactive rheostatic response using a pathogen injection that induces fever? There was an upward shift in both the core body temperature (i.e., fever) and gloved hand temperature preference! The shift at the higher- and lower-end of the core body temperature indicates that a new set point was established for the duration of the fever response (Figure 6.5). Crucially,

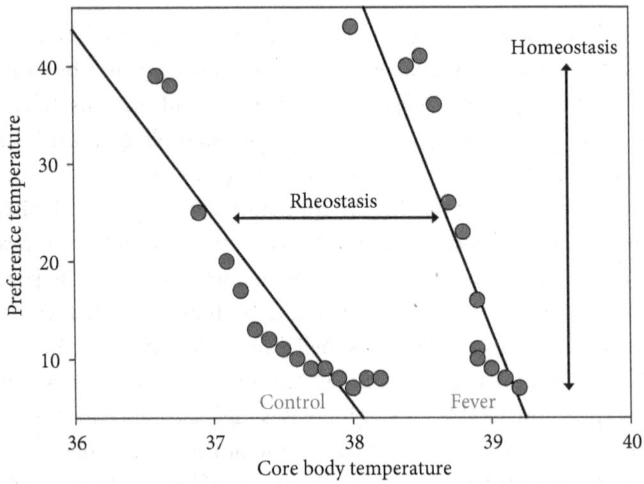

Figure 6.5 Dissociation of reactive rheostasis and homeostasis control of core body temperature.
An inverse (i.e., negative) relationship is maintained between core body temperature and preference for skin temperature in human subjects. Subjects administered an attenuated pathogen increased core body temperature and exhibited fever symptoms. The rightward shift in the line indicates a new set point is established and is a reflection of the reactive rheostatic mechanisms that are engaged. Note that the negative relationship between core body temperature and preference temperature is maintained in both control and fever states, evidence that homeostatic mechanisms are held constant at both high and low ends of the resting and fever adjusted core body temperatures.
Data are adapted from Massonnet and Cabanac, 1974, and used with permission from John Wiley and Sons.

homeostatic responses are still maintained despite the increased set point. Homeostatic control of body temperature is illustrated by the persistent negative relationship between core body temperature and preferred temperature of the skin in the gloved hand. The maintenance of the negative relationship (i.e., homeostasis) despite the increased set point (i.e., rheostasis) provides clear evidence that these two physiological systems can easily be dissociated.

Theoretical framework linking homeostatic core body temperature and the rheostatic fever response

Given that the control of body temperature can be dissociated into homeostatic and rheostatic mechanisms, how do these two systems interact with the models proposed in previous chapters? As fever induces an increased set point (i.e., 40°C), a reactive rheostatic model must act on the hypothalamic centers to control the elevated core body temperature for the entire bout (Figure 6.6). The homeostatic sensor of core body temperature includes temperature receptors and temperature channels located in many visceral organs and the brain. Any change in core body temperature must be integrated by error detectors that involve the interaction between warm- and cold-sensitive neurons in the preoptic area of the anterior hypothalamus. Homeostasis under healthy conditions is maintained at a set point of 37°C, however during the onset and maintenance of fever, a new set point is established at 40°C. This increased core body temperature is triggered by the release of PGE2 from brain endothelial cells that are near circumventricular organs. The sensors for the reactive response start with the receptor for interleukin-1β on hypothalamic endothelial cells, which serve as error detectors and respond to increased circulating interleukin-1β by producing PGE2. PGE2 provides an evolutionary ancient signal that elevates core body temperature and maintains a new, and temporarily increased, set point established by temperature-sensitive neurons in the preoptic area. The change in temperature-sensitive neurons in the preoptic area is then communicated to the controller (i.e., dorsomedial hypothalamus), which provides the same function for reactive rheostatic and homeostatic control of core body temperature. The difference between fever-induced and healthy core body temperature is the effector tissues that serve to maintain a higher temperature than the common 37°C set point. Under the circumstances of regulated increase in core body temperature, effector responses include the increased

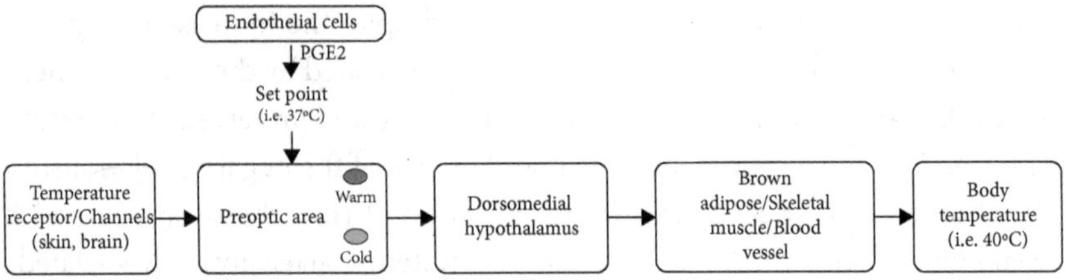

Figure 6.6 Schematic representation of a reactive rheostasis model for fever. Core body temperature homeostasis is maintained by temperature-sensitive receptors/channels that act as sensors for changes in external temperature (via skin) and internal temperatures (i.e., brain/viscera). Deviation in core body temperature is detected by neurons in the preoptic area that include both warm-sensitive (red) and cold-sensitive (blue) neurons. These neurons function to compare internal temperature changes with the set point of approximately 37°C. An error signal is conveyed from the preoptic area to the dorsomedial hypothalamus, which acts to control the correction in core body temperature by signaling via the autonomic nervous system to increase shivering by skeletal muscles, constriction of blood vessels, and thermogenesis by brown adipose tissue. The initiation of reactive rheostasis originates with the detection of cytokines by brain endothelial cells. Here, PGE2 overrides the set point to establish a higher threshold of 40°C. The reactive rheostatic control of fever is maintained in the presence of continued production of prostaglandin by endothelial cells. The absence of PGE2 results in a return to homeostasis and body temperature regulation around 37°C.

thermogenesis by brown adipose tissue, muscle induced shivering, and vasoconstriction of the circulatory system. Once the levels of PGE2 decrease, the fever set point is lost, and the homeostatic regulation of body temperature around 37°C is re-established. The regulated change in core body temperature is then switched accordingly, and the negative feedback control is reinstated.

Summary

In this chapter a comparative perspective on fever-induced regulation of body temperature was used to present an example of a single bout of reactive rheostasis. As the innate immune response comprises genetically coded material, it should not be surprising that most of the key molecules that form the primary physiological defense are evolutionary conserved and ancient. The reactive rheostatic mechanisms involved in the control of fever-induced regulation of core body temperature consist of specific genes (i.e., PGE2) and cellular mechanisms (i.e., endothelial cells) that act on existing homeostatic pathways. The key genes that separate rheostasis from homeostasis here include the production of PGE2 and cyclooxygenase 2. The production of PGE2 is dependent on exposure to cytokine produced by the immune cells (i.e., macrophages). A common pattern has emerged in which a reactive rheostasis mechanism induces slight modifications to the set point and consequently error-detector that lasts a relatively long (i.e., 5–9 days) defined period. Now, we will expand on the concept of reactive rheostasis by looking at regulated changes in physiology over much longer timescales, for example, pregnancy.

Note

1. A Nobel Prize anecdote: One rule in awarding the prize is that the recipient must be living. Unfortunately, Ralph Steinman had died the week before the announcement, the news of which had not reached the committee, and he did not personally receive the award. He remains the only person to be "mistakenly" awarded the prize posthumously!

7

The reactive response of life

Mothers are all slightly insane

J. D. Salinger (1919–2010), *The catcher in the rye*

The protagonist in *The catcher in the rye*, Holden Caulfield, travels to New York by train and comments to himself that "mothers are all slightly insane." The remark stems from a conversation with Mrs. Morrow, mother of fellow classmate Ernst Morrow, during which she lies about her son's popularity and humility. The quote reflects Holden's opinion that mothers irrationally believe only the good, however untrue, about their children, and (in general) express an unconditional love toward their offspring. Arguably the impact of children on the psychological and physiological health and wellness of their parents is one of the most understudied of biological phenomena. This is evidenced by the amount of research in the PubMed.gov database, which contains 5 times the number of publications for "fetal development" compared to "mother and pregnancy." Perhaps Holden's comment might well be rephrased "Mothers are all slightly scientifically underappreciated." In this chapter, the focus is on the regulated physiological changes a mother will experience during pregnancy, starting with the response to the fertilization of her egg (or specifically ovum) until the birth of the child.

In Chapter 6, we covered a relatively short-term (i.e., 5–7 days) reactive rheostatic fever response. When considering pregnancy, we will predominantly use human evidence due to the larger and detailed scientific literature relating to it. The aim is to highlight that the mother mounts a reactive rheostatic response to the fertilization of her egg that involves a complex, yet timed series of physiological changes, which are independent of homeostatic responses. It is important to note that reactive rheostasis is not a response that occurs in the developing embryo or fetus as these represent developmental stages that are an entirely separate physiological process. The history

On Rheostasis. Tyler John Stevenson, Oxford University Press. © Oxford University Press 2024.
DOI: 10.1093/oso/9780197665572.003.0007

of the study of fetal development is beautifully described in the wonderful book *A history of embryology* by Joseph Needham (1959), while *The dance of life* (2020) by Professor Magdalena Zernicka-Goetz and Roger Highfield provides a modern insight on how human life forms.

Across the animal kingdom, fetal development can be subdivided into two types that each require regulated physiological changes in the mother, and in few cases the father. The most common type of fetal development is *oviparous*, but *viviparity* is the predominant form in mammals. The words used to describe each type of fetal development are derived from the Latin root word "parus," which means "bearing." The key difference is the addition of "ovo" or "vivus" which correspond to "egg" and "alive" respectively. By using these two simple words most of the animals in the world can be classified as either oviparous, producing eggs, or viviparous, producing live offspring. Rarely, there also ovoviviparity, such as in some frogs and seahorses. After internal fertilization, the embryos develop inside eggs and do not have a physiological connection to the parent. The animals hatch from the egg and are born "alive" resembling the combination of *ovo* and *vivi* parous forms. From an evolutionary perspective, oviparous animals are believed to be the ancestral form (Stewart, 1997).

Oviparity is common in birds, most reptiles, and some fishes; in mammals it is restricted to the monotremes (i.e., platypus). Oviparity can be further subdivided into (true) oviparity— the predominant form—in which animals engage in copulatory behavior and fertilization of the egg occurs internally and ovuliparity in which the female lays unfertilized eggs that are externally fertilized (common among most fishes and frogs). Animals in which the development of the fetus occurs within the parent which subsequently gives birth to live free-living offspring, are called viviparous (Blackburn, 2015). These viviparous animals, as we will discuss below, experience a series of reactive physiological responses with predictable well-defined regulated changes across a set period commonly referred to as gestation. A series of genetic, molecular, and physiological changes must occur for the mother to ensure her own survival and that of her developing fetuses during gestation. Similar processes occur in oviparous animals that incubate their eggs, but here the reactive rheostatic response includes egg laying and tactile feedback during brooding. In birds, brooding behavior includes the motivation to sit on the eggs, which ensures an optimal incubation temperature for fetal development. But as we will see, the tactile stimuli received from the eggs has a powerful impact on the brooding parents.

Comparative analyses of length of the period for fetal development

Both viviparity and oviparity have defined periods, the length of which varies considerably across the animal kingdom but deviates very little within species. Gestation in mammals, or the time from fertilization to birth, has a wide range of periods (Figure 7.1). Across mammals, there is a strong correlation between gestational period and the size of the offspring, and in particular the size of the brain (Sacher and Staffeldt, 1974). In most animals in the order rodentia (e.g., mice, rats) the length of gestation is roughly 3 weeks. For humans, the during of gestation is typically 38–40 weeks. In most species in the order Equidae, the duration of gestation is almost an entire year. Compared to the African bush elephant (*Loxodonta africana*), these gestational periods are relatively short, as a female elephant is pregnant for almost 2 years! Marsupials have an alternate strategy in which the mother can engage in a phenomenon known as embryonic diapause and keep a fertilized egg in a state of dormancy until the environmental conditions are optimal for the offspring's survival (Renfree and Shaw, 2000). In kangaroos, the duration of gestation averages 5 weeks; however, if the female already has a joey in

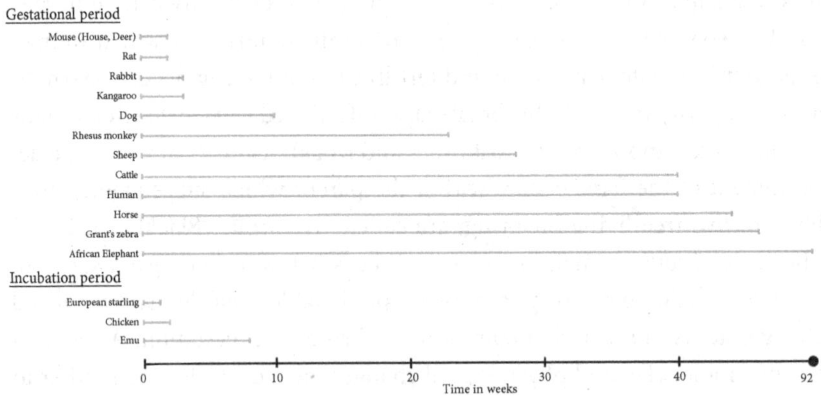

Figure 7.1 Comparative perspective of gestation and incubation. Gestation varies widely across mammals with a relative short duration for rodentia (i.e., mouse, rats) and longer duration for Equidae (i.e., horse, zebra). African elephants have the longest gestation duration of an average 92 weeks. Incubation length in birds has less variability and can be as short as 2 weeks in Passeriformes (e.g., European starling) or up to 8 weeks in Casuariiformes (i.e., Emu).

her pouch, she can exhibit embryonic diapause in which the early embryonic cells, called a blastocyst, stop multiplying and instead remain in an arrested state. This control over embryonic development permits optimal timing of multiple births that is entirely dependent on nutritional and environmental conditions.

Oviparity can be associated with incubation in birds, which includes the time from egg laying to hatching of the juveniles. Typically, incubation is significantly shorter than mammalian gestation. For birds in the order Passeriformes—*perching* birds such as the European starling (*Sturnus vulgaris*)—the length of egg incubation is typical 2–3 weeks. In the most abundant bird species, the domestic chicken, incubation is similar in length to gestation in rodents, approximately 3 weeks. The bird species with the longest known incubation is the Wandering albatross (*Diomedea exulans*), at approximately 85 days (Tickell, 1968). Most bird species show high levels of biparental care (Cockburn, 2006). Male and female birds generally taking an active role during incubation and show a strong motivation for brooding behavior. Recent evidence has indicated that approximately 81% of bird species form strong pair bonds and biparental care (Cockburn, 2006). Biparental care observed in birds provides a very interesting opportunity to identify sex-specific and sex-independent reactive rheostatic responses. Comparisons across species show that even though the duration of gestation and incubation varies widely across animals, there are evolutionary conserved sequences of genetic, molecular, physiological, and anatomical stages that underpin the energetic demands required within both processes to support reproduction. Let's start where it all begins: fertilization.

In the absence of sex, many reproductively competent females will cycle through the programmed rheostatic change described in Chapter 4. Recall that during the follicular phase of the mammalian reproductive cycle, the corpus luteum produces the hormone progesterone. In the absence of fertilization by a sperm, the corpus luteum ceases to produce progesterone and develops into the corpus albicans and is degraded. When fertilization occurs, the combination of one set of chromosomes from each parent leads to the formation of a zygote. It is at this very moment that an entirely innate reactive physiological response occurs, that is independent from the programmed rheostatic reproductive cycle. This response within the mother will trigger a series of well-orchestrated physiological responses that will forever change her physiology, immune function, and often behavioral and cognitive states. Oviparous animals go through a different sequence of events that will be described later in the chapter. A key point to note is that regardless of whether animals are oviparous or viviparous there are defined phases in the process of gestation and egg laying/incubation that can be explained by rheostatic processes. Regardless of reproductive strategy, the reactive rheostatic processes invoked include a discrete set of newly formed hormones and anatomical structures that have direct impacts on the mother (and in some cases the father). That gestation and egg laying/incubation is a rheostatic process will be demonstrated by describing the parallel stable physiological control of energy availability.

Cellular and hormonal basis of viviparity

After fertilization of the ovum by a sperm, a single-celled *zygote* is formed. The cell then starts to divide into two, then four, then eight to create a *morula* (from the Latin for mulberry) within 3 days, consisting of 16 cells that form a spherical structure. Each cell within the morula has the remarkable capacity to become any tissue cell in the organism, an ability referred to as totipotency (Posfai et al., 2021). As the cells continue to divide, a blastocyst is formed, characterized by a fluid filled cavity inside the dividing cells giving rise to an "outer" and an "inner" cell mass (Figure 7.2). The inner space of the blastocyst becomes the blastacoel cavity and a cluster of cells is seen which is the inner mass. The inner cell mass will become the embryoblast and ultimately form the embryo and subsequently, the baby. The blastocyst is a critical stage of development as it is here the cells of the inner cell mass start to differentiate into specialized populations (Shahbazi and Zernicka-Goetz, 2018). In kangaroos this is the stage at which embryonic diapause can occur and development arrested. The cells that surround the inner cell mass and encapsulate the cavity are called trophoblasts. They are a specialized groups of cells that will form the placenta and produce two key hormones that drive reactive physiological responses in the mother. The blastocyst consists of around 200–300 cells, and its formation after fertilization takes appropriately 7 days. The stage for the initial sensory detection to trigger the reactive rheostatic response is set and requires the implantation of the blastocyst into the mother's uterus. The uterus acts as the physiological sensor, and the action of progesterone signaling produced by the corpus luteum has served to prepare the uterine lining for the implantation of the blastocyst (Wetendorf and DeMayo, 2012).

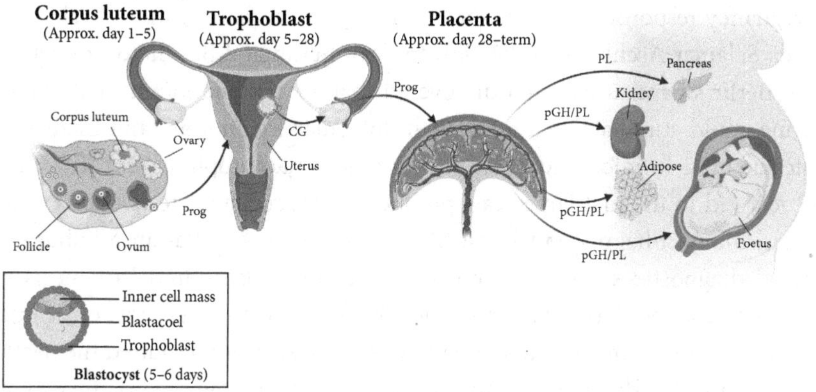

Figure 7.2 Anatomical and hormonal steps involved in reactive rheostasis during pregnancy.
Prior to fertilization of the ovum by sperm, the corpus luteum secretes progesterone (Prog), which prepares the uterus for implantation by the blastocyst. Trophoblast cells release chorionic gonadotrophin (CG) to stimulate the ovaries to produce additional progesterone. Ovarian derived progesterone acts to facilitate placental development and trigger the production of placental lactogen (PL). As the placenta develops, it continues to release placental lactogen and placental growth hormone (pGH), which act on multiple tissues, particularly those involved in energy homeostasis (e.g., muscle adipose and liver).

The process of implantation of the blastocyst also includes the production by the trophoblast cells of an entirely new hormone: chorionic gonadotropin (Rothbauer et al., 2017). As detection of chorionic gonadotropin forms the basis of pregnancy tests, it is likely one of the most important and commonly tested hormones in human populations. Chorionic gonadotropin was discovered over a century ago by Otfried Fellner at the Institut für Allgemeine und Experimentelle Pathologie at the University of Vienna (Fellner, 1912). He showed that injections of pieces of human placenta into immature rabbits resulted in growth of the uterine and vaginal epithelium, signs of early pregnancy. This work led Otfried to conclude that a biochemical substance (i.e., hormone) produced by placental tissue was responsible for preparing the mother's uterus for pregnancy! Subsequently, Toyoichi Hirose at the Osaka Medical College confirmed that repeated injection of extracts of human placental tissue stimulated ovulation and normal luteal function in immature rabbits (1920). The precise hormone released by the placenta to trigger an early pregnancy response in female mammals was not isolated until several years later. Selamar Aschheim at the University of Berlin and Bernhard Zondek at the Berlin-Spandau Hospital discovered that a hormone found in women's urine could stimulate ovarian growth, follicular maturation, and increased uterine mass in mice, and called it chorionic gonadotropin (1928). This discovery led to the first large-scale production of pregnancy tests. For several years, the Aschheim-Zondek test (A-Z test) was provided as an Edinburgh-based diagnostic service for clinicians and hospitals around Britain, as it could be used to determine pregnancy in women with a very low error rate (<2%). However, the test took 5 days, was expensive, and required the death of a mouse. The drive for quicker pregnancy tests led Maurice Friedman and Maxwell Lapham at the University of Pennsylvania to develop a similar technique using rabbits, and they were able to identify growth of the rabbit ovary after only one day (this became known as the rabbit test). The quicker results obtained using rabbits instead of mice meant the rabbit test soon replaced the A-Z test (1931). Unfortunately, the test still required the examination of the ovaries post-mortem, and every test resulted in the death of an animal. Both of these tests have now been superseded by tests provided by the local pharmacy, which continue to assay the presence of chorionic gonadotropin in a woman's urine sample but do so without the need for loss of animal life.

The primary physiological function of chorionic gonadotropin is to stimulate the corpus luteum in the ovary to maintain high levels of circulating progesterone (Figure 7.2), which serves to prepare and synchronize the function of multiple organs and the placenta for pregnancy. Within the placenta, chorionic gonadotropin is important for establishment of the syncytiotrophoblasts, a specialized layer of "embryonic" cells that are in direct contact with the maternal blood supply and function to regulate maternal-embryo nutrient exchange (Huppertz, 2008).

Once the blastocyst has implanted into the uterus, gastrulation occurs. In humans this occurs roughly 3 weeks post-fertilization and is characterized by the formation of three layers of cells in the developing embryo: the outer ectoderm, the middle mesoderm, and the inner endoderm (Solnica-Krezel and Sepich, 2012). These layers will develop into the (1) skin, hair, brain, spinal cord, and peripheral nervous system, (2) bones, muscle, kidney, gonads, and circulatory system, and (3) the stomach, colon, liver, pancreas, bladder, and lung, respectively. This is accompanied by expansion of the trophoblast cells, which start to form the placenta, and this also provides a critical part of the reactive rheostatic response of the mother.

The placenta: a master physiological controller of gestation

During pregnancy the placenta grows in proportion to the developing fetus and the production of the hormone placental lactogen, which provides a reliable indicator of pregnancy "time" and its presence codes for the length of pregnancy. By 24 weeks of gestation, the placenta mass is approximately 250 mg, but it will continue to grow and will reach a mass of over 700 mg by term (Thompson et al., 2007). The placenta has an amazing, layered structure across the entire animal kingdom, but has a conserved function from fish to humans. Across mammals, the placenta shows structural diversity, which has made comparative studies and the development of translational animal models extremely tricky for the biomedical sciences. One of the authoritative literature sources on placenta structure and function is *Comparative placentation* (2008) by Peter Wooding and Graham Burton. Graham Burton is now Emeritus Mary Marshall and Arthur Walton Professor of Physiology of Reproduction and was director of the Centre for Trophoblast Research located in the Department of Physiology, Development and Neuroscience at the University of Cambridge. To highlight the wide-ranging complexity in variation of placental structure, a few examples of other species are provided next.

The two main aspects of the placenta that differ across species are (1) the shape and distribution of connections between the fetal membranes and maternal endometrium and (2) the layers of tissue between the fetal and maternal vascular systems (Wooding and Burton, 2008). The placental shape in humans, as in most primates, is *discoid.* A discoid placenta also develops in rabbits, rodents, and bats. Dogs, cats, seals, bears, and elephants have zonary placentas, in which a complete or incomplete band of tissue surrounds the fetus, while horses and pigs have a diffuse placenta, in which almost all of the allantochorion forms the placenta. A fourth placental type, cotyledonary, is found in many ruminant species such as cows and goats. Here multiple, discrete areas of attachment called cotyledons are formed between the allantochorion and the endometrium. Specifically, fetal cotyledons contact maternal caruncles, the cotyledon-caruncle complex being a placentome.

Across all mammals, six membrane layers are formed during the establishment of the placenta. Three of these layers are derived from fetal extra-embryonic membranes and are retained irrespective of placental type: (1) fetal endothelial cells, (2) fetal connective tissues and (3) chorionic epithelial cells. In humans, maternal blood has direct contact with the three fetal tissues, as all maternal layers (i.e., endometrial epithelium, connective tissue, and uterine endothelium) are degraded. The loss of all three layers is described as hemochorial and is also found in most rodents. In other mammals, such as horses, pigs, and sheep, all three maternal layers are retained, and the placenta is referred to as epitheliochorial, while in others, such as dogs and cats, there is a mix of degradation and retention and only the uterine endothelium is maintained, and the layered structure is called endotheliochorial.

The placenta functions to remove fetal waste products, but crucially provides oxygen and nutrients to the embryo and serves to maintain energy stability for both mother and fetus. The primary means by which the placenta communicates nutrient requirements and demands is via the production of placental lactogen (Figure 7.2). During pregnancy, placental lactogen is secreted by the placenta and acts to bind to receptors for growth hormone and prolactin. Its primary function is the regulation of maternal lipid and carbohydrate metabolism. However, placental lactogen release does not occur until the second and third trimester and peaks shortly before birth. Placental lactogen functions similar to growth hormone and regulates energy balance in the mother thereby ensuring there are ample nutrient supplies for the developing fetus. The increase in placental lactogen produced during pregnancy provides a clear hormonal signal that is unique for the regulated change in physiology required for pregnancy and a key driver for the reactive rheostatic response for maternal energy balance.

Although a transient organ that is generally considered "clinical waste" in most westernized countries, the placenta has strong spiritual meaning for many cultures. Across cultures, there are considerable differences in the value placed on the placenta. The most widespread method to discard the placenta is a ceremonial burial. As many cultures consider the placenta a "guardian angel" for the baby, the practice of burying the placenta under a tree is common. In Bali, a placenta may be buried with a coconut. Among the Maori people of New Zealand, the word for placenta and land is the same, *whenua*, and the organ is buried on tribal land. Similarly, among the Navajo of the United States, the placenta is buried at the boundary of the family land to connect the child's spirit with the ancestors. In Hmong culture, the placenta is considered the child's "jacket" and provides a spiritual link that helps to bring the child into the world and to move into the afterlife. For some cultures, the umbilical cord is not severed, and the placenta remains connected to the baby, until after 3–10 days the cord dries and breaks off. The rationale for this *lotus birth* varies, but the prevailing view is of a spiritual relationship between the baby and placenta (Buckley, 2003). Most mammals engage in placentophagy, or the act of eating or consuming the placenta to get nutrition. Placentophagy is extremely rare in human cultures, but in some traditional Chinese medicines, dried human placenta known as Ziheche is used to treat wasting disease, infertility, and impotence. Please note that there is no scientifically proven benefit, medical or otherwise, to eating a placenta, dried or wet (Farr et al., 2018).

An overview of the finely tuned hormonal changes that underlie the reactive rheostatic changes in placental and fetal development is provided in Figure 7.3. Early in pregnancy, chorionic gonadotropin concentrations rise gradually, which is associated with a small transient increase in circulating progesterone concentrations. The increase in chorionic gonadotropin provides an early cue to signal the presence of a developing embryo and prepare the mother for upcoming nutrient demands. Then there is a massive increase in placental lactogen and progesterone concentrations begin to gradually increase to match the increased size of the placenta and fetus. During the second and third trimester, the nutrient demands of the fetus are immense and placental lactogen concentrations rise to provide the long-term rheostatic signal. This signal is required to control the regulated changes in physiology that are needed to ensure sufficient energy stores are available for both the mother and fetus, and to buffer against any short-term homeostatic changes in energy balance.

Homeostatic and reactive rheostatic demands of energy balance during pregnancy

Now that the main cellular, hormonal, and anatomical substrates of reactive rheostasis are covered, let's focus on how regulated physiological responses can be separated from homeostatic stability during pregnancy. One of the best studied physiological responses in the mother during pregnancy is energy stability, and it provides a clear example to highlight the key differences between homeostatic and rheostatic physiological systems. Producing an offspring requires massive energetic demands. The Mayo Clinic, an internationally recognized source for medical recommendations, suggests that a healthy woman with a BMI of 18.5–25 should typically gain 25–35 pounds (11–16 kilograms) during pregnancy. Most of the weight gain will be attributed to

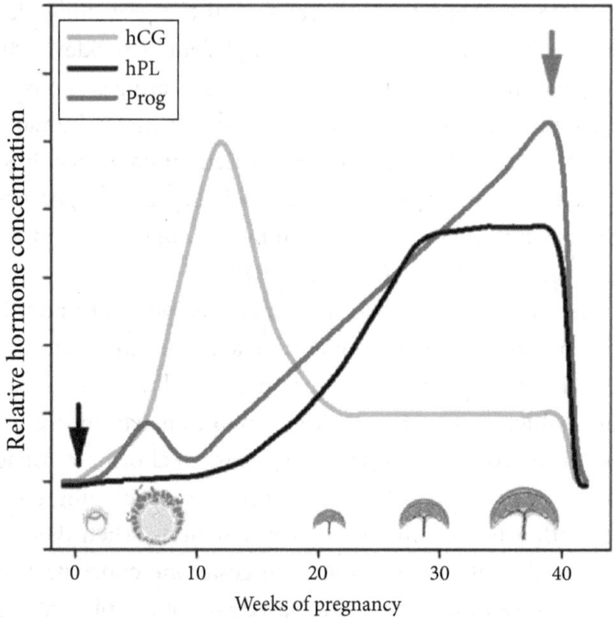

Figure 7.3 Regulated changes in placental and ovarian hormones during pregnancy.
Several hormones show predictable and timed increases in blood concentrations during key pregnancy phases, including placental hormones chorionic gonadotropins (hCG) and placental lactogen (hPL). CG peaks early in pregnancy while levels of PL increase gradually from mid-gestation. Maternally derived hormones that originate from the ovary, including progesterone (Prog), function to support the developing embryo and display a steady increase in blood concentrations for the duration of pregnancy. Approximate development stage of trophoblast and placenta are provided above the x-axis. Size of cell/tissue is a guide and not to scale.

the developing embryo and placenta and associated wet weight (e.g., blood). The mother therefore must change her energy status so that in addition to maintaining normal physiological functions she can also support the growth of the fetus and its life support systems. The regulated changes in energy balance that are seen during gestation are evolutionarily conserved.

If we consider the changes in human body mass and glucose tolerance during gestation, we can clearly see separate reactive rheostatic and homeostatic regulation occurring at hormonal and molecular levels. For the first half of the pregnancy (i.e., 20 weeks) the levels of insulin required to keep glucose levels within the normal range remains relatively similar to the nonpregnant state. After 20 weeks' gestation, there is a steady increase in the amount of insulin required to regulate blood glucose levels, it being doubled by week 30 week and tripled prior to term. Immediately after delivery of the offspring, however, the concentrations of insulin required to maintain normal blood glucose concentrations rapidly and substantially declines to pre-pregnancy levels (Marcinkevage and Venkat Narayan, 2011). Given the changes in glucose regulation during pregnancy, women will typically have their glucose metabolism checked during the second trimester to screen for gestational diabetes, high levels of insulin (i.e., hyperinsulinemia), and the risk of excessive glucose (i.e., hyperglycemia) and fetal overgrowth. This is commonly done by means of an oral glucose tolerance test. A blood sample is taken in the morning after an overnight fast; the woman is given 50 g of a glucose solution and a second blood sample is taken after 2 hours to determine how the body metabolizes sugar. A healthy glucose tolerance test will result in blood sugar levels that are lower than 155 mg/dl (or 8.6 mmol/L). In the event a woman has values higher than 155 mg/dl, a second test may be administered to determine the risk for gestational diabetes (Forest et al., 1983). A women's response to an oral glucose test is remarkably consistent across pregnancy, with a very strong correlation between tests early in pregnancy and late pregnancy in the absence of gestational diabetes. So, a women's response to the 50 g glucose solution in the first trimester will be nearly identical when tested again in the third trimester (Nahum and Huffaker, 1990). But what does this mean in relation to the naturally occurring increase in insulin during pregnancy? The consistency in oral glucose tests across time indicates that the homeostatic response to glucose challenge is maintained and the innate, first-order stimulus-physiological response remains unchanged. Yet, the reactive rheostatic regulated change in insulin concentrations is controlled over the long term to ensure adequate nutrient availability for the mother and fetus. Any deviation from the reactive rheostatic response, such as in women who are obese or underweight, has detrimental effects on the metabolic health of both mother and offspring (Lawlor et al., 2012).

So, what underlies the reactive rheostatic response in glucose regulation during pregnancy? In effect, women will develop insulin resistance, a physiological process in which the adipose tissue, liver, and muscles reduce their responsiveness to insulin signaling and stop using glucose from the circulatory system for energy (Catalano, 2002). Under these physiological conditions, of continued high levels of circulating glucose, the pancreas makes more insulin.

During pregnancy, placental lactogen drives the reactive rheostatic activation of the pancreatic islets, which enhances insulin secretion (Brelje et al., 1993). Placental lactogen drives increased insulin production and secretion by stimulating the expression of genes involved in energy balance, such as homeobox 1 in pancreatic and duodenal tissue (Lombardo et al., 2011). One hypothesis is that placental lactogen shifts the maternal physiological state to elevated insulin during mid to late pregnancy and promotes the availability of glucose and amino acids to the fetus (Handwerger and Freemark, 2000). The mother, then, will rely on other stored nutrients, such as fatty acids and ketone bodies, to ensure short-term homeostatic energy balance (Butte, 2000). The effect of elevated insulin levels on peripheral tissues and the maternal central nervous system, especially the brain, is poorly understood (Ladyman and Brooks, 2021).[1] While placental lactogen is the primary driver in stimulating the increase in insulin production during pregnancy, another placental growth hormone is also produced during gestation; it is secreted at a constant level early during gestation but starts to increase at approximately 20 weeks, coincident with the increase in circulating insulin concentrations (Anthony et al., 1998). The increase in placental growth hormone has multiple effects across peripheral tissues. In muscles, placental growth hormone acts to decrease the activity of a key gene in the insulin pathway—the insulin receptor substrate-1 associated PI 3-kinase (Barbour et al., 2004). This prevents insulin driven glucose adsorption by muscle cells and in turn results in higher circulating glucose. Together, placental lactogen and placental growth hormone serve to induce the reactive rheostatic change in energy balance by initiating increased insulin by the mother and control resource allocation in the former, and ensure maternal tissues are protected from elevated insulin by inducing resistance in the latter. Therefore, there is a coordinated effort by placental hormones to control the reactive rheostatic stability through complex, transient restructuring of peripheral tissues and the subsequent change in regulated physiological variables (e.g., insulin).

Framework for the reactive rheostatic response during pregnancy

Now we can identify the physiological steps, starting with ovulation and ending with full-term pregnancy, and identify the key components involved in the reactive rheostatic response. During the menstrual (or estrous) cycle, the unfertilized ovum is released during ovulation and the corpus luteum produces a higher concentration of the hormone progesterone. Corpus luteal derived progesterone acts on the uterus to prepare the tissue for implantation of a fertilized cell mass (i.e., zygote). Here, progesterone facilitates decidualization, endometrial receptivity, and embryo implantation. After implantation, the production of chorionic gonadotropin by the trophoblasts signals to the ovaries to continue to produce progesterone, which facilitates fetal and placental development. After 10 weeks, the placenta starts to synthesize and secrete progesterone and becomes its major source, with levels increasing during pregnancy until birth. As the placenta expands, placental lactogen and growth hormone coordinate long-term changes in energy balance (i.e., glucose) with direct impacts on the mother's body mass.

The reactive rheostatic model described here for the physiological changes during pregnancy follows a similar framework to the programmed rheostasis that was described in Chapters 3, 4, and 5. The only difference is that it lacks the feedback/feedforward input to the sensor (Figure 7.4). In this model reactive rheostatic system, the sensory detector resides in the uterus and is driven by implantation of the blastocyst. Increased production of chorionic gonadotropin by the trophoblasts stimulates the error detector, the ovary, to increase synthesis of progesterone. Here a new regulated set point for circulating progesterone is maintained by the ovary. Increased progesterone secretion acts on the developing placenta (i.e., former trophoblast cells), triggers

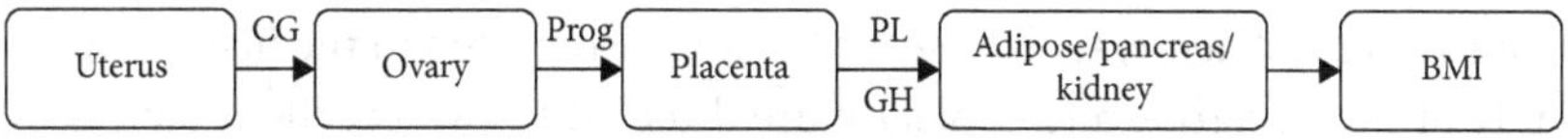

Figure 7.4 Reactive rheostasis model of mammalian pregnancy.
In a reactive rheostasis model, the uterus is the primary tissue that acts as the detector of implantation of the blastocyst and initiates the coordinated physiological response. Implantation and subsequent production of chorionic gonadotrophin (CG) by the trophoblasts is conveyed to the ovary. Here, the ovary acts as the error detector and begins to produce significantly more progesterone (Prog) to support the growth of the placenta and foetus. The placenta is the master controller of pregnancy and functions as the interface for energy balance between mother and foetus. The release of placental hormones lactogen (PL) and growth hormone (GH) orchestrates the long-term, regulated change in peripheral tissue function during pregnancy. Peripheral tissues (e.g., pancreas) act as effectors during pregnancy and serve to ensure the timed responsive changes in the regulated variable, such as body mass index (BMI).

growth, and ensures the development of the fetus. The placenta, then acts as the controller to support fetal development, but also produces placental lactogen and growth hormone. Placental lactogen acts in multiple tissues involved in energy homeostasis (i.e., pancreas) and coordinates trade-offs in nutrient demands between the mother and fetus. The coordinated actions of placental lactogen and growth hormone regulate the function of effector tissues that in turn regulates long-term changes in mother energy balance (e.g., body mass index).

While we have focused on human pregnancy, the reader should be aware that a historical problem exists when we consider hormones in that many were named based on an associated function, which often than not this relates to mammals, even though the same hormone can have multiple or different functions in the same or other species. For example, progesterone was named for the ability to stimulate the proliferation of the uterine endometrium layer in gonadectomized adult rabbits, independent of another ovarian hormone, estrogen (Corner and Allen, 1929). Similarly, prolactin is commonly associated with the promotion of lactation and generally assumed to be associated with milk production from breast tissue. The discovery of prolactin, however, was entirely due to a research program that used pigeons and a peculiar anatomical structure in birds, the crop sac. In the 1930s, Oscar Riddle and colleagues demonstrated that injections of the pituitary gland into pigeons resulted in the production of a milk substance in the crops sac (Riddle et al., 1933). This milk substance has a similar biological function to breast milk, in that juvenile birds will feed on it until they are self-sufficient in eating other nutrients, such as seeds. Readers interested in the history of the study of progesterone and prolactin are directed to two excellent papers by Ali and colleagues (Ali et al., 2020) and Smith (2004).

Cellular, hormonal, and behavioral basis of reactive rheostasis in oviparous species

Despite the clear structural differences between mammalian gestation and egg-laying in birds, there is a remarkable similarity in the hormonal regulation of the reactive rheostatic response. Both of the hormones estrogen and progesterone are essential to support egg-laying, despite their misnomer. In birds the initiation of egg laying and the incubation of eggs otherwise called brooding is dependent on progesterone and an additional hormone prolactin, interestingly, as fetal development often occurs outside of the body as described earlier in this chapter. Birds show a very high level of bi-parental care and the incubation of the eggs until hatching is an affair that often involves both mother and father. Here we will focus on the hormonal, anatomical, and sex-specific physiological responses in birds during the process of fertilization, egg formation, incubation, and ultimately, hatching of the offspring.

The timing of ovulation in birds is remarkably like that in mammals, despite the fact that birds often lay a clutch of eggs, and highlights the evolutionary conserved nature of vertebrate reproduction. Birds experience high levels of gonadotropin-releasing hormone and luteinizing hormone production prior to ovulation. Luteinizing hormone stimulates the development of follicles and increased synthesis of estrogen, the hormone critical for the creation of most egg-yolk components, and stimulates oviductal growth (Johnson, 1986). This growth establishes a hierarchy in which the largest follicles are selected to release ova into the oviduct (Etches and Petitte, 1990). Concurrently, luteinizing hormone is also involved in stimulating the production of progesterone in the granulosa layer of the follicles. Interestingly, these steroidogenic effects are not matched, as estrogen production is greatest in the small follicles and lowest in the large yolky follicles (Johnson, 1986), whereas progesterone synthesis is greatest in the late-stage large yolky follicles. The increase in progesterone and estrogen production by the follicles is essential for the preparation of the oviduct for the release of an ovum and its maturation into fully formed eggs (Figure 7.5), a process parallel to uterine growth in mammals.

Clearly there are key structural and functional differences in birds, compared to mammals, which mean that in birds blastocyst implantation into the uterine lining does not occur. Instead, the follicle releases the ovum directly into the oviduct, which consists of a series of regions: the infundibulum, magnum, isthmus, and finally the shell gland (i.e., uterus). For birds, the infundibulum is the location in which the ovum is fertilized by the sperm, a process that occurs relatively quickly (i.e., seconds–minutes). After fertilization, the zygote descends into the magnum where the different egg-white proteins are stored, albumen encases the cells, and starts to form an egg, which will reside in this structure for approximately 3 hours. Unlike mammals in which the fetus gains nutrients from the maternal blood supply, the egg white encasing the chick embryo has a high protein content and is the source of nutrients for *in ovo* growth. From the magnum, the egg moves

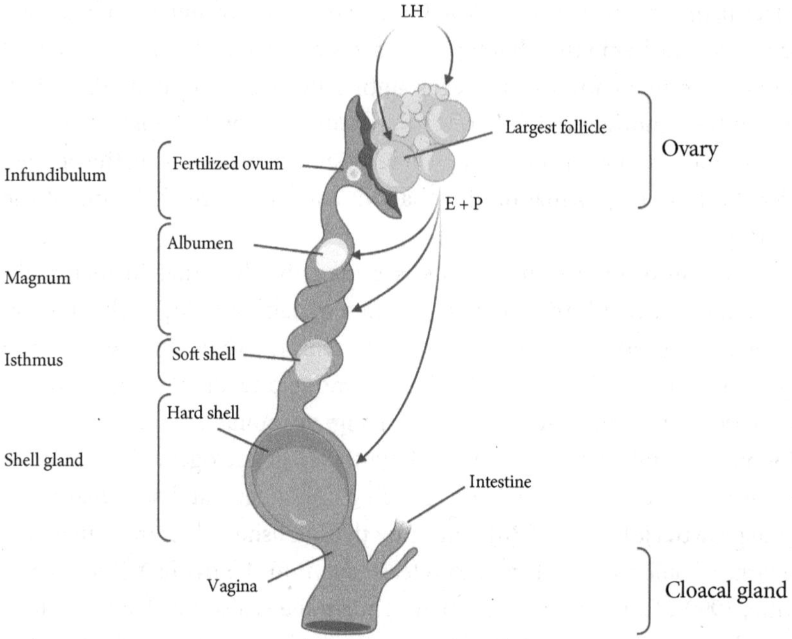

Figure 7.5 Structure and hormonal signaling during the timing of egg development

The avian female reproductive tract includes the ovary, oviduct, vagina, and cloacal gland. Luteinizing hormone (LH) secreted by the pituitary gland stimulates the synthesis of estrogen (E) in small follicles and progesterone (P) in large follicles. Estrogen and progesterone act to facilitate fertilization of the ovum in the infundibulum and stimulate oviduct growth. In the oviduct, the fertilized ovum travels from the magnum into the isthmus and then finally into the shell gland. In the magnum, albumen is formed over a period of 3 hours, after which a soft shell is established in the isthmus over 1–1.5 hours; the outer hard shell of the egg develops in the shell gland over a period of 21 hours. The mature egg is then passed into the vagina before released out of the cloacal gland.

into the isthmus where it remains for 1–2 hours, during which time the protective eggshell membranes are formed around the albumen, which will support mineralization and eventual production of the shell. The solid eggshell is made during the 18–22 hours the egg remains inside the uterus (also called the shell gland). After complete mineralization of the eggshell (calcite crystals), the egg is moved briefly (~1hr) into the vagina before laying. While the egg is in the reproductive tract there can be bi-directional exchange of hormones or any chemical signal between the mother and the developing embryo, but once the egg is laid, this form of communication ceases. These well-timed stages serve to ensure that reproduction in birds occurs with remarkable precision, and the entire length of time from ovulation to egg laying is approximately 24–26 hours.

When the egg is laid, the embryo is still in early development and must be kept at an appropriate temperature, etc. This is achieved through incubation by adults. The early ethologist, Daniel Lehrman, working at Rutgers State University in New Jersey, identified key behavioral and hormonal stages of incubation behavior in one of the best studied birds for incubation behavior, the ring dove (*Streptopelia risoria*). Like most birds, both male and females will take turns incubating the eggs until hatching approximately 2–3 weeks later (Lehrman, 1964). Gregory Ball examined the timing of incubation behavior in male and female rings doves when he was a graduate student in the laboratory of Professor Rae Silver at Columbia University. They engage in a well-timed exchange of incubation behavior in which the males sit on the eggs for a duration of time in the middle of the day and the female will incubate for the rest (Ball and Silver, 1983). Strikingly, non-incubating bird initiate the exchange in behavior, which revealed a strong motivation for the partner to sit on the eggs and not because the other bird became lazy and opted to negate parental duties. These findings indicated a strong internal drive to engage in parental behavior that requires a complex synchronization between birds as well as intermate physiological synchrony!

But what is the hormonal driver of brooding behavior? Prior to ovulation, gonadal steroids are high, and this primes the mother to express incubation behavior (Cheng and silver 1975). Courtship behavior also leads to accumulated exposure of the male to progesterone, which is critical for the induction of brooding in male ring doves (Cheng, 1975). After a female hen lays an egg, there is a rapid inhibition of the reproductive system, which is characterized by a decrease in gonadotropin-releasing hormone, luteinizing hormone, and follicle-stimulating hormone (Dunn et al., 1996). These subsequently result in lower circulating levels of estrogen and progesterone. The drop in gonadal steroids however means that incubation behaviors will not necessarily be maintained. Key to completion of the reactive rheostatic process of egg-laying and brooding behavior, needed for successful reproduction in birds, is touch feedback from the eggs. This is clearly shown by the fact that in the absence of eggs, achieved by simply removing eggs from the nest for a few hours, brooding behavior in both sexes significantly decreases (El Halawani et al.; 1980, Goldsmith et al., 1984). In chickens, the hen will continue to lay eggs in the absence of a sufficient accumulation (i.e., 12 eggs) to obtain a full "clutch"—a behavioral trait that has been capitalized effectively by poultry farmers for millennia.

Once incubation has started, there is a significant rise in prolactin, and this ensures the maintenance of brooding behavior (Buntin, 1986). The tactile stimulation from the eggs stimulates more prolactin synthesis and release into the bloodstream. It is this long-term increase in prolactin release that is essential for the maintenance of brooding behavior in birds (Lehrman and Brody, 1964), the control of which is entirely driven by tactile feedback from eggs (Buntin, 1986). Interestingly, gonadotropin-releasing hormone and luteinizing hormone concentrations can be reversed within 4 days by simply removing the eggs from an incubating female (Dunn et al., 1996). Similarly, removal of the eggs or complete nest will induce a substantial decrease in prolactin concentrations and can be reversed by simply returning the eggs in turkey hens and canaries (El Halawani et al., 1980; Goldsmith et al., 1984, respectively).

The increase in prolactin secretion that results from tactile stimulation from the eggs is facilitated by the development of a "brood patch" in birds—a featherless surface on the underside or abdomen of birds. This anatomical region becomes highly vascularized in the time leading to egg-laying and functions to increase the area of contact between the bird and eggs in order to facilitate heat transfer (Lack, 1968; Murton and Westwood, 1977). A contemporary of Daniel Lehrman, Robert Hinde, professor of zoology at the University of Cambridge, was also a pioneer in the study of behavioral endocrinology of birds. Using canaries, Hinde demonstrated that the brood patch developed prior to egg-laying and was characterized by loss of feathers, vascularization, and thickening of the abdomen (Hinde, 1962). In collaboration with Elizabeth Steel, Hinde showed that estrogen produced by the ovaries caused the vascularization of the brood patch (Steel and Hinde, 1964). What is more, the loss of feathers from the brood patch was also in part caused by estrogen, but another hormone, such as prolactin, is required for complete loss of the brood patch feathers. While brood patch formation in females can be seen to be driven by hormonal changes associated with egg formation and laying, brooding behavior is also observed in male birds. The next question is what hormonal factors could be involved in stimulating brood patch development? Phalarope birds have a sex-reversed parental role in which the males build the nest and brood the eggs. In both male Red-necked phalarope (*Phalaropus lobatus*) and Wilson's phalarope (*Phalaropus tricolor*), testosterone injections coupled with prolactin are sufficient to induce defeatherization and increased vascularity of the brood patch (Johns and Pfeiffer, 1963). These data establish that gonadal hormones in both male and female birds are essential for the formation of the brood patch. The development of the brood patch provides a vital role for the long-term reactive rheostatic regulation of parental physiology and behavior in birds.

Framework for the reactive rheostatic response of egg laying and brooding

A rheostatic model of incubation in birds must include mechanisms that are present in both males and females due to the observed bi-parental behavior. This negates the use of any feedback signals that are derived from the egg during the passage through the oviduct. Instead, the complete model, from sensor detector, error detector, set point, controller, to effector for a regulated variable, in this case brooding, must be assigned to an evolutionary conserved brain and peripheral structures shared between both sexes (Figure 7.6). The initial stage involves an anticipatory production of steroid hormones associated with breeding activity by the ovary and testes. In males the increase in testosterone and in females the increase in progesterone and estrogen by large and small follicles stimulates brood patch development. Instead of implantation, which acted as the sensor signal for mammals, in birds the skin of the abdomen, the brood patch, contains receptors that are hormonally primed and sensitive to tactile signals from the eggs that initiate the reactive rheostatic response. The brood patch, then, acts as the error detector. This structure transmits tactile information into the central nervous system and triggers a neuroendocrine response. Here, dopamine neurons in the arcuate nucleus of the hypothalamus are activated and stimulate the release of prolactin from the pituitary gland. Arcuate nucleus dopaminergic neurons provide the control for the reactive rheostatic response by maintaining high prolactin release in the presence of continued sensory input from the eggs. Elevated prolactin acts on multiple tissues (e.g., abdomen, adipose tissue) to maintain brooding behavior for the duration of the incubation phase. The reactive rheostatic control of parental behavior continues for the duration of egg incubation by a reinforced brood patch–contact prolactin brooding

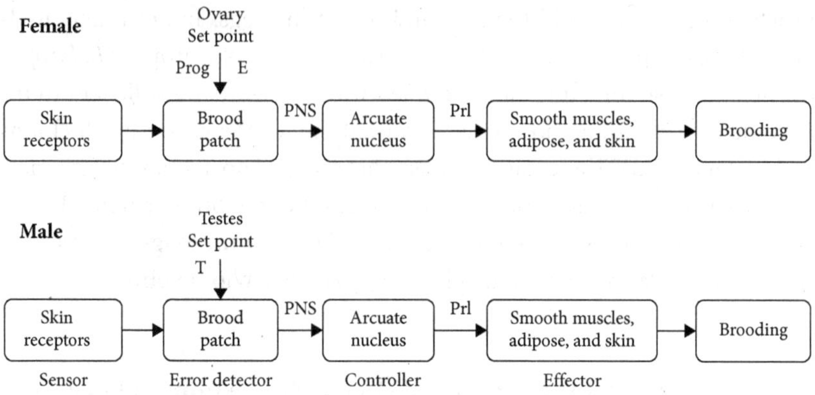

Figure 7.6 Reactive rheostasis model for brooding in birds.
The sensor to initiate the reactive rheostatic response is triggered by touch receptors (i.e., mechanosensitive) located in the abdomen. During egg formation gonadal steroids including estrogen (E), progesterone (Prog), and testosterone (T) signal a new set point that triggers the development of the brood patch, which functions as an error detector. Tactile signals from eggs are then transmitted via the peripheral nervous system (PNS) into the hypothalamus. Dopamine cells in the arcuate nucleus/infundibular region of the hypothalamus act as the controller and regulate the release of prolactin (Prl) from the pituitary gland. Increased prolactin release acts in the brain and a range of peripheral tissues (e.g., smooth muscles, brood patch) to maintain brooding behavior until the eggs hatch.

behavior response. The termination of the response develops in the absence of eggs either due to hatching or removal by predators.

Summary

This chapter has focused on another type of reactive rheostasis and included examples to highlight that well-defined, sequential regulated change in physiology underlies gestation in mammals and brooding in birds. We have clearly seen that female energy balance during pregnancy is maintained through short-term homeostatic regulation of glucose. The long-term change in insulin action during pregnancy, however, clearly illustrates a regulated change that is driven by the coordinated activity of an entirely separate genetic (i.e., placental lactogen and growth hormone) and anatomically (i.e., placenta) driven response. Using brooding physiology and behavior in birds, evidence was presented to demonstrate that the reactive rheostatic mechanism for parental behavior need not be sex-specific and instead can be activated in both mother and father. These reactive rheostatic responses, unlike programmed rheostasis, have a defined period which can vary from a few weeks (i.e., chicken) to several months (e.g., African elephants). A limitation of the concepts presented here is that there is considerable variation in parental physiology and behavior across the animal kingdom. The variation makes it challenging if not impossible to formulate a single overarching model that can encompass the reactive rheostatic regulation of embryo development. In Chapter 8, we will return to the concept of allostasis and use the stress response to describe the hierarchical organization of physiological stability. The chapter will focus on the regulated variable, stress hormones (e.g., glucocorticoids) and illustrate that homeostatic, rheostatic and allostatic processes are separate mechanisms that function independently to control an individual organism's response to an environmental stressor.

Note

1. Dr. Jodi Pawluski is a Canadian Behavioral Neuroendocrinologist and a member of the RPC-C Canadian Professional Counsellors Association. Dr. Pawluski's podcast Mommy Brain Revisited is an excellent resource to learn more about the impact of the embryo and child on the mother's brain, see jodipawluski.com/mommybrainrevisited.

8
Hierarchical organization of physiological stability

Everyone knows what stress is, but nobody really knows.
János Hugo Bruno Seyle (1907–1982).

Stress is a complex physiological, behavioral, and cognitive response. For all vertebrates, the physiological basis of stress responses is natural and occurs as a reaction to an event that can be real or perceived, or in anticipation to an event. Stressful events can be physical (e.g., injury) or psychological (e.g., exams, losing or starting a job). For most vertebrates, what is determined as a stress event(s) may not be consistently observed across all organisms, but primarily depends on the individual. During development, most animals mount a physiological reaction to stressful events, such as the first-time prey encounter a potential predator. A *stressor* can be anything causes a stress response and can be either a perceived threat or danger and real event. As an individual gains experience with a particular stressor, the ability to anticipate stressful events can change the physiological basis of the stress response. This anticipatory change in the physiological response is different from the programmed rheostatic responses we have discussed so far, as it is dependent on an individual's prior experience and involves cognitive and behavioral processes. The ability to modify physiological responses through anticipation requires a series of brain regions that control learning, memory, and personality. In this regard, the amygdala and hippocampal structures are involved in learning and memory and have bidirectional connections with brain regions involved in personality, such as the frontal area.[1] Amygdala-hippocampal-prefrontal cortex activity is the primary reason for such diverse individual variation in human stress responses, for example, exams are more stressful for some people while driving a car may be less stressful for others. One reason that studying stress in animals, including humans, is

On Rheostasis. Tyler John Stevenson, Oxford University Press. © Oxford University Press 2024.
DOI: 10.1093/oso/9780197665572.003.0008

tricky, stems from the fact that a single stressor can induce a range of different stress responses, the proportion and magnitude of which can vary between individuals. This has led to the common recognition of being *stressed* but not fully understanding how the diverse range of individual stress responses to the same stressor is controlled.

As discussed in Chapter 1, most physiological responses involve homeostatic and rheostatic mechanisms that are driven by innate programs. Here we will primarily focus on the concept of allostasis, or the ability to maintain physiological stability through change (Sterling and Eyer, 1988). The term allostasis is derived from the Greek *állo* or "different; other" and is a conceptual model that provides a predictive error or preventative response that generally enhances survival and fitness. The concept of allostasis was also developed, in part, to help account for the vast range of individual differences in a regulated physiological response within a population. The stress response is one physiological process that provides a clear example of allostasis. Unlike homeostasis and rheostasis, allostasis does not have a defined set of genes that have evolved to maintain physiological stability, but instead is regulated by the central nervous system via higher-order brain regions. This chapter will first describe the anatomical structures and hormones involved in the stress response and then outline the physiological changes that occur in response to a stress event. Next, the range of allostatic responses will be highlighted with an emphasis on the neuroanatomical basis of the stress response. The chapter will conclude by returning to the conceptual framework of how homeostasis and rheostasis function, how they relate to the stress response, and will incorporate how allostatic mechanisms can affect the key neural and cellular substrates. The overall aim of the chapter is to illustrate a hierarchical organization of physiological stability that involves primary (i.e., homeostasis), secondary (i.e., rheostasis), and tertiary (i.e., allostasis) levels.

The adrenal gland: hormone producing tissue

The effector of the stress response is a small piece of tissue located on top of the kidney and called the adrenal gland. Its name is derived from Latin: *ad* which means of or relating to and *renal* referring to the kidney. The adrenal glands are imbedded within nephric fat rostral to the kidneys and are predominantly involved in the synthesis and release of several hormones. Because the adrenal glands are surrounded by large depots of fat tissue, early anatomists did not clearly identify them until Bartolomeo Eustachius finally discovered and described them in the 1550s (Miller, 2013). In 1805, Jean Léopold Nicolas Frédéric Cuvier, a French zoologist identified two distinct anatomical features of the adrenal gland: a medulla and a cortex region (Figure 8.1). That the tissue was essential for life was subsequently characterized by Charles-Edouard Brown-Sequard (1817–1894) in a series of studies in which he removed the adrenal gland, a procedure known as adrenalectomy, and discovered that it often led to death (Brown-Sequard 1856). But because patients and animals often died from adrenalectomy, the function of the adrenal gland remained a mystery for most of the nineteenth century. After his death in 1895, Giuseppe Marzo was discovered to have an enlarged adrenal gland, a condition called congenital hyperplasia. Marzo had a complicated life as he had been classified as a female at birth, but by the age of 4 years he was redesignated as a male (de Creccio, 1865). As an adult, Marzo had male secondary sexual characteristics (e.g., penis, facial hair) but

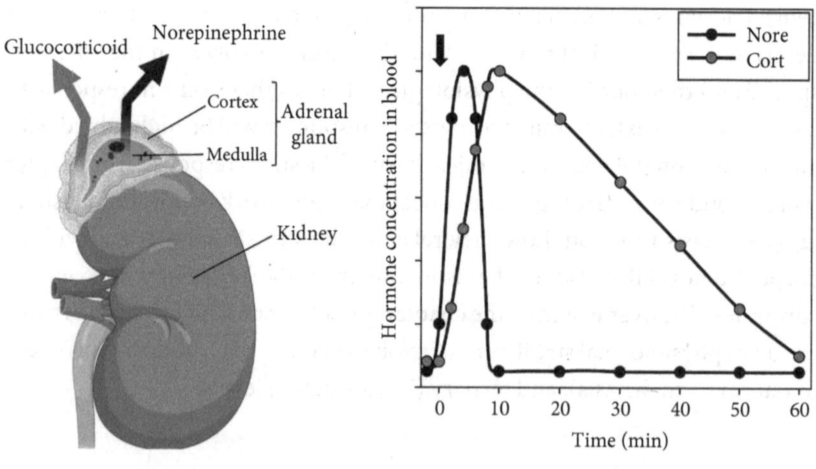

Figure 8.1 Adrenal gland anatomy and the physiological response to a stressor. Two hormones are primarily involved in the stress response. The rapid release of the neurotransmitter norepinephrine (Nore) from the adrenal medulla serves to quickly allocate energy resources to heightened arousal and engage muscles for quick action. The second response is slower acting and of longer duration and includes the release of glucocorticoid (i.e., cortisol (Cort)) from the adrenal cortex. The release of norepinephrine occurs as a bolus release and lasts a several minutes. Glucocorticoid release occurs over a longer period (e.g., 30–60 minutes) and then returns to the homeostatic set point concentrations.

autopsy of the cadaver revealed that Marzo had female reproductive tissues including a vagina, uterus, fallopian tubes, and ovaries (de Creccio, 1865). Cases such as Marzo's resulted in it being proposed that the function of the adrenal gland was to regulate growth, blood pressure, and reproduction, but the exact hormones through which it achieved such control remained elusive. It was not until the twentieth century that physiologists started to discover that the cortex and medulla produced different classes of hormones.

That the adrenal (or suprarenal) glands contained a substance with dramatic pharmacological effects was first shown in 1893 by George Oliver, a Harrogate physician, and Edward Schäfer, professor of physiology at University College London. However, a name was not coined for the substance until John Abel in the United States prepared crude adrenal extracts in 1897 and called them epinephrine. In 1901, Jokichi Takamine (1854–1922) an American Japanese biochemist working in his Clifton, New Jersey, laboratory was the first to isolate, characterize, and patent epinephrine from the adrenal medulla. It is now established that the adrenal medulla secretes two hormones: epinephrine and norepinephrine.

Edward Calvin Kendall (1886–1972), Philip Showalter Hench (1896–1965) and Tadeusz Reichstein (1897–1996) discovered that the adrenal cortex produced several hormones, such as cortisone. All three shared the 1950 Nobel Prize for Physiology or Medicine for their discoveries of adrenal cortex hormones, their structure, and biological effects. The primary hormones synthesized and secreted by the adrenal cortex include: (1) mineralocorticoids that are required for salt and water balance; (2) glucocorticoids, which are required for the anti-inflammatory response during illness, the regulation of aspects of metabolism, and glucose production; and (3) androgens (e.g., testosterone), which are sex hormones produced in both male and females.[2] The release of hormones from the adrenal medulla occurs from chromatin cells, which are modified neurons and are controlled by the sympathetic nervous system and called the sympatho-adrenal-medullary system. Hormones synthesized in the adrenal cortex are under the neuroendocrine control of neurons located in the hypothalamus, a pathway known as the hypothalamo-pituitary-adrenal axis. Here, corticotropin neurons in the paraventricular nucleus (PVN) trigger the release of adrenocorticotropin from the anterior pituitary gland, which travels via the circulatory system to the adrenal cortex to initiate the synthesis and release of glucocorticoids. In reptiles, birds, and most rodents, the key glucocorticoid hormone is corticosterone; in primates, the primary hormone is called cortisol. The different names are due to the slight differences in the chemical structure of the hormone, however, the function in the regulation of metabolism, immune reactions, and the stress response is the same.

The physiological basis of the stress response

The physiological response to a stressor is consistent across vertebrates: epinephrine is rapidly released from the adrenal medulla to quickly activate energy stores to initiate the "fight or flight" response (Figure 8.1). This was first coined by Walter B Cannon to describe the physiological reaction to perceived (real or unreal) stress events, produced as a reaction to many life-threatening situations, such as needing to fight off a predator or running to a safe location. It is characterized by increased heart rate/force of contraction, respiration, perspiration, and blood glucose due to the breakdown of glycogen by the liver as well as decreased gastrointestinal activity. The rapid increase in epinephrine/norepinephrine secretion from the sympathetic post-ganglionic nerves and the adrenal medulla activates these physiological changes, but also acts in the brain to induce a heightened level of cognitive arousal to the perceived danger. However, the fight or flight response does not conform to models of homeostasis proposed by Cannon. Despite identifiable controllers (e.g., sympathetic nervous system) and regulated variable (i.e., epinephrine), there is no well-defined error detector or set point to maintain peripheral levels of epinephrine/norepinephrine. Crucially, the rapid action of epinephrine/norepinephrine could not account for prolonged stress responses. Therefore, other mechanisms must be recruited to account for the additional long-term components of the stress response. It was not until several years later when one of the pioneers of the field of stress research, János Hugo Bruno (or "Hans") Seyle, a Hungarian Canadian endocrinologist and professor at the Université de Montréal. Seyle (1946), proposed the General Adaptation Syndrome, conceptually defined as "the sum of all *non-specific*, systemic reaction of the body which ensue upon long continued exposure to stress. It is distinct from the specific adaptive reactions, such as the development of the musculature following prolonged physical exercise, the allergic and immunologic phenomena elicited by foreign proteins or micro-organisms."

Seyle's model for the general adaptation syndrome consisted of an initial *alarm* reaction that required a shock, then counter shock steps, which were then followed by the stages of *resistance* and *exhaustion*. The alarm reaction was the non-specific response to a wide range of stress events that can induce what was referred to as a "call to arms" or more commonly associated with the fight or flight of the body's defenses. This early phase is associated with increased release of epinephrine from the adrenal medulla. The second stage of the response, or Resistance, is associated with an increased release of adrenocorticotrophic hormone from the anterior pituitary gland, an enlarged adrenal cortex, and elevated secretion of glucocorticoids (i.e., cortisol/corticosterone) from the adrenal cortex (Figure 8.1). During the resistance stage an adaptation to the same alarming stimulus develops and a form of habituation occurs. However, if prolonged exposure is continued the adaptative response deteriorates and pathological conditions from the elevated action of glucocorticoids manifest in disease conditions. In the event a lack of adaptation occurs during the resistance, a final stage of exhaustion develops and ends in the death of the animals. The key developments of Seyle's model are the stages of resistance and exhaustion.

How can we study the physiological mechanisms induced during the human stress response? Several types of experiments, and natural conditions, induce a stress response. One common experimental test in human research uses the Trier Social Stress Test developed at the University of Trier to examine a moderate level of stress in human participants (Kirschbaum et al., 1993). It runs as follows: there is an initial baseline rest and task instructions, in which the participants wait in a room for 30 minutes. A saliva and/or blood sample is collected and used to determine the typical resting levels of many dependent variables, such as cortisol and adrenocorticotropic hormone. Then participants are given the task instructions that consists of a role-playing game in which the participants have a couple minutes to prepare a 5–10-minute speech to a panel of investigators and convince the members that they are a perfect candidate for a job. The subjects are then given a final cognitive task that usually involves simple mathematics. A second saliva and/or blood sample is collected immediately prior to the stress procedure, and a further sample is collected immediately after the math questions. There is a final recovery period in which the participant returns to the resting room and saliva and/or blood samples are collected every 15–20 minutes for up to 2 hours (Allen et al., 2014). Human baseline blood concentrations are typically 30 pg/mL epinephrine, 1 ng/mL norepinephrine, 50 pg/mL adrenocorticotropin and 200 ng/mL cortisol. The Trier Social Stress Test results in all hormones showing an approximate doubling in blood concentrations that can last for several minutes to hours. The peak of epinephrine/norepinephrine release from the sympatho-adrenal-medullary system typically occurs within 5–10 minutes of the onset of the stressor and is followed by an equally rapid decline in blood hormone concentrations. However, the hypothalamic-pituitary-adrenal axis regulated release of cortisol is not seen to begin until 10–15 minutes after the onset of the stress event, and hormone concentrations can remain elevated for an extended period of time.

The sympatho-adrenal-medullary system and hypothalamo-pituitary-adrenal axis pathways are both recruited during the initial stress response; however, repeated presentation of the same stressor usually results in an attenuated physiological response, a process known as habituation. In participants that engage in the Trier Social Stress Test, the concentrations of cortisol and adrenocorticotropin are significantly reduced in some individuals after 8 attempts of presenting their case for the job candidate (Gerra et al., 2001). The reduction in hormone release indicates that the central nervous system can adapt its response to the stressful event and that individuals can gain a level of control over their physiological response. In other participants the concentrations of cortisol and adrenocorticotropic hormone remained elevated after the test, which would suggest that there was a level of individual variation in the ability to habituate to a social stressor. This variability in the cortisol response contrasts with epinephrine release, which does not habituate to repeated stressors (Gerra et al., 2001), and highlights how the sympatho-adrenal-medullary system pathway constantly responds to any perceived stressor, whereas the response of the hypothalamo-pituitary-adrenal axis can be modified with experience, leading to adaptative responses to the same stress event.

A series of tests have been developed for rodents to assess their physiological responses to physical and psychological stressors, immobilization being the most widely applied stressor. The procedure is simple and involves placing a rat or mouse into a restrainer (i.e., ventilated tube) for 30 minutes or more. One of the strengths of this procedure is the ability to conduct repeated immobilization tests over a short period of time to examine the impact of chronic stress and habituation in rodent physiological stress responses. Rats that experience an immobilization test show an increase in circulating norepinephrine, adrenocorticotropic hormone, and glucocorticoid (i.e., corticosterone) production. Repeated immobilization tests results in habituation evidenced by a decline in glucocorticoid production with repetition (Pitman et al., 1988).

Allostasis is an individual specific physiological response

Not all stressors induce the same response equally across organisms or even individuals. For most birds and mammals, the ability to anticipate and prepare for a subsequent stressor provides strong adaptive value. This leads to challenges in defining what determines if something is stressful. An individual's cognitive state and behavior play an essential role in the perception of and response to any single event. Allostatic responses permit a change in physiological stability, for example a dampening of a stress response to a particular event can be adaptive if there is no real danger or if the situation is not life-threatening, which prevents physiological "over reactions" or wasted behavior. Unlike some types of homeostatic control, allostasis does not require feedback and instead forms an "open-loop" system. Bruce McEwen (1938–2020) proposed three characteristics that determine how an individual will respond to a stress event (McEwen, 2000): First, the way in which the individual interprets the stressor specifically—or how the brain determines whether an event is a threat or something to be avoided. Second, the ability to alter the physiological response to perceived stressors can be affected by exercise and healthy diet (Tsatsoulis and Fountoulakis, 2006). Finally, genetic factors. If one monozygotic twin is diagnosed with major affective disorder, the other twin is are over 60% more likely to show major affective disorder. For fraternal (i.e., dizygotic twins) the likelihood of the second twin being diagnosed with major affective disorder drops to 20% (Plomin et al., 1994). These three characteristics of allostasis permit adaptive responses and facilitate the fitness of an individual over the short term.

Allostatic load refers to the cumulative effects of stress or a lifestyle that must cope with continuous stressful events. Long-term exposure to physical, emotional, and environmental stressors induces a cost that can lead to chronic pathology and death. Four types of allostatic load have been characterized and are depicted in Figure 8.2. In the first example, stress events induce the typical robust change in glucocorticoid concentrations (i.e., physiological response) but an individual does not show the adaptive "resistance" stage, and a higher hormonal concentration is maintained for a prolonged period. The second type of allostatic load is a lack of physiological response to a stress event and is demonstrated by a modest increase in glucocorticoid concentration that quickly subsides into a muted response. In cases such as these there is an inadequate physiological response to address the prevailing stress event. The third type of allostatic load is a lack of habituation to a constant stressor

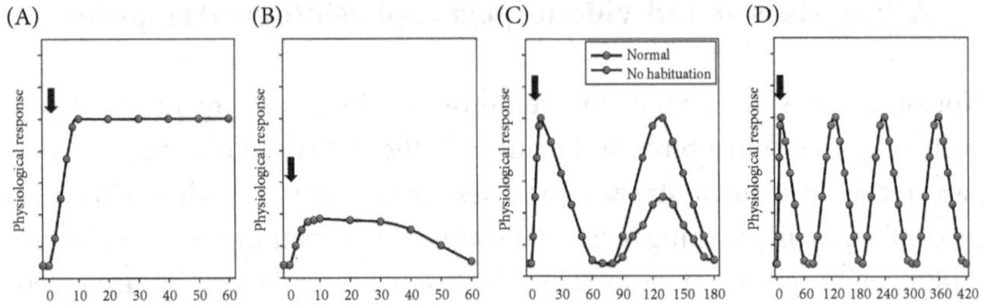

Figure 8.2 Allostasis and stability in the stress response.
There are 4 types of allostatic load that include (A) a sustained elevation in the physiological response, (B) the lack of a physiological response, (C) the absence of habituation or adaptive physiological response, and (4) exposure to multiple different stressors in successive events. Black arrows indicate the presence of a stressor.

Source: Data are adapted from McEwen 1998 and used with permission from the New England Journal of Medicine.

and an individual shows repeated physiological stress response to the same cue. The loss of adaptation to a constant stimulus results in high glucocorticoid concentrations circulating in the blood. The final type of allostatic load is when an individual is exposed to multiple different stress events at a relatively high frequency. The repeated "hits" of physical, emotional, or environmental cues cause heightened glucocorticoid concentrations that reduces an individual's fitness (McEwen, 1998). The control of allostasis and allostatic load is achieved by the brain, and as indicated above, involves the amygdala, hippocampus, and frontal cortex.

A series of interconnected brain regions govern individual stress responses

Studying how the brain controls stress responses in animals faces many challenges. Ethologist such as Nicolas Tinbergen and Konrad Lorenz recognized the importance of studying animal behavior and physiology from the perspective of how the animal senses and integrates the environment, and how the animal's behavior could alter the perception of the environment. This approach was heavily influenced by Jakob Johann Von Uexküll (1864–1944), a German biologist who founded the Institut für Umweltforschung at the University of Hamburg. Von Uexküll described the sensory representation of the animal's world as the merkorgan and that of an animal's behavior and its behavioral effect on the environment as the werkorgan. These two components are critical for the umwelt, or the "self-centered world." The combination of the sensor (merkorgan) and effector (werkorgan) actions formed a feedback loop that produced the individual's umwelt (Von Uexküll, 2010 [1934]). Jakob proposed that different organisms can have different umwelts, even though the animals share the same environment. A series of anatomical regions in the brain form functional connections that act to provide the individual animals with their umwelt. Brain regions such as the cortex (or pallial structures in non-mammals) are involved in establishing how the environment is represented. The parietal lobe includes several cortical regions involved in how a mammal perceives touch, smells, taste, sounds, and vision, which form part of the merkorgan. These cortical regions have direct connections with limbic regions or parts of the brain that control the emotional processes and memories of an individual. The frontal cortex, hippocampus, and amygdala structures are three of the predominant brain regions that provide mammals with personality, memory, and emotional processing. In the case of allostasis, the prefrontal cortex, amygdala, and hippocampus form an interconnected circuit that codes an individual's perceptual representation of the world and environmental stimuli (Radley et al., 2015). These brain regions are fundamental to how allostatic responses integrate prior experience with a range of environmental stimuli (e.g., social, physical, perceptual) and individual experiences (i.e., genes, development) to modify the stress response (Figure 8.3).

Major technological advances during the 1990s have allowed scientists to image the human brain in real time. By using a series of standardized tests, researchers can now develop a brain *atlas* that is a visual representation of an individual's cognitive and mental representations. Crucial to the discussion in this chapter is the ability to create visual images of brain regions that are active during the stress response and determine the association of increased or decreased brain activity with physiological responses, such as stress-induced increases in glucocorticoids (i.e., cortisol). The Montreal Imaging Stress Task is a modification of the Trier Social Stress Test that was adapted specifically for brain imaging studies. In this test, participants are placed in a brain imaging device such as a functional magnetic resonance imaging (fMRI) system that allows for the measurement of changes in brain activity (i.e., changes in localized blood flow). During the "rest" baseline conditions, participants view a static computer screen while laying inside the

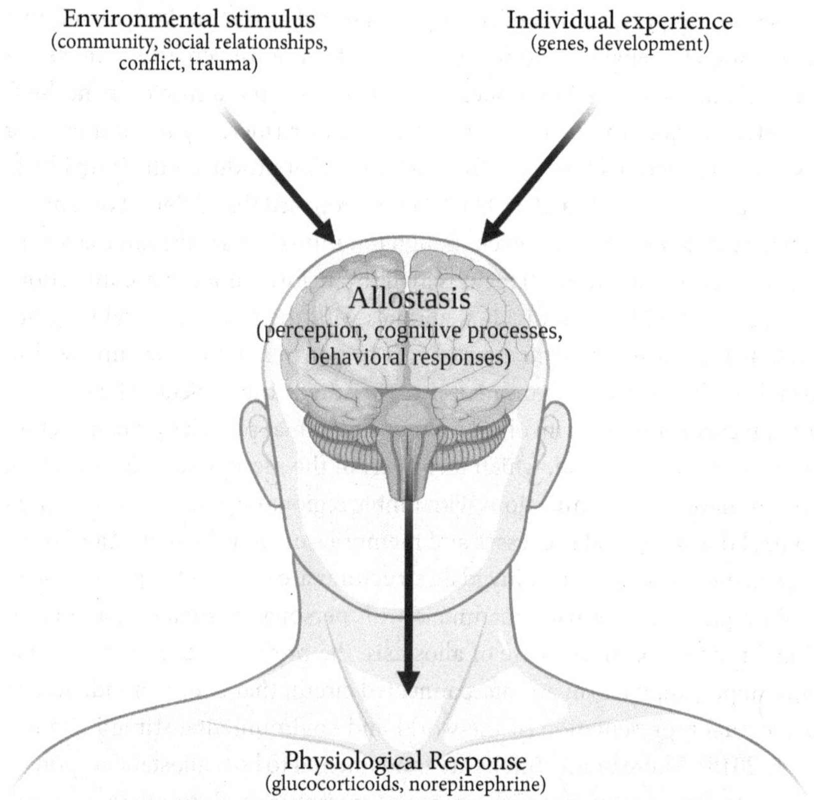

Figure 8.3 Environmental and individual differences in allostasis. Long-term physiological responses to stress events or stressors such as the local community or social relationships are perceived differently across individuals. This variation is in part determined by an individual's developmental or prior experiences. The combined early life experiences and exposure to stress events shape how an individual will perceive and subsequently form cognitive and behavioral responses that impact homeostatic and rheostatic mechanisms. Higher-order brain regions such as the prefrontal cortex, amygdala, and hippocampus interact with the hypothalamus to induce long-term changes in physiological variables and glucocorticoid release.

magnetic resonance imager. Then, a series of control visual displays and tasks are presented that include simple mathematical questions, and changes in brain activity are recorded. The key 'stress condition' that is tested consists of advanced mathematical questions that the subject must complete in a very short period. The result is then compared which an "expected" performance

(Dedovic et al., 2005). The expected performance value is based on the scores achieved by other participants, to provide a "social evaluation" stress event. Participants that complete the Montreal Imaging Stress Test show significant increases in circulating cortisol concentrations during the stress conditions compared to both control and resting states, which is consistent with physiological responses to the Trier Social Stress Test. By comparing brain activity between the simple math and the stress condition (timed advanced mathematical questions), researchers can identify changes in brain activity associated with stressors. In the Montreal Imaging Stress Test, fMRI patterns in individuals that produced a stress response (i.e., "responders") showed a significant reduction in brain activity in the frontal cortex (i.e., medial orbitofrontal cortex) and hippocampus (Dedovic et al., 2009). Similar reductions in frontal cortex, hippocampal, and hypothalamic activity have also been observed using another brain imaging technique, positron emission tomography, where a strong inverse correlation between brain region deactivation during the stress condition and stress-induced increase in cortisol was also observed (Pruessner et al., 2008). These data highlight that, in humans, the frontal cortex and hippocampal regions show acute activation in response to stressors. These short-term responses have significant adaptive value, as acute exposure to stress-induced increases in glucocorticoids can have a positive effect to increase arousal and allocation of energy resources to facilitate behavioral and cognitive performance and fitness.

How are these brain regions linked to the loss of allostatic regulation of stress responses? Long-term exposure to stress events or the development of allostatic load leads to maladaptive coping mechanisms due to altered brain activity. Several studies have shown that the density or expression of glucocorticoid receptors throughout the limbic forebrain are reduced in response to stress (Sapolsky et al., 1984a; Furay et al., 2008). Changes in the distribution and abundance of glucocorticoid receptors impact the negative feedback regulation of the hypothalamo-pituitary-adrenal axis and the capacity of cortical, hippocampal, and amygdala circuits to support adaptive cognitive and behavioral change (Dallman and Jones, 1973). Glucocorticoid receptors in the hippocampus are one of the key mechanisms that provide allostatic control over the stress response. Crucially, the glucocorticoid receptor expressing cells in the hippocampus are critical for stopping the increased glucocorticoid production by the adrenal cortex during the acute stress response (Sapolsky et al., 1984b). A loss of receptors is directly attributable to elevated and prolonged production of glucocorticoids and is a type of maladaptive allostatic responses. An excellent review on the role of the amygdala, hippocampus, and prefrontal cortex in the modulation of stress responses is provided by Radley and colleagues (2015).

Evidence for a strong link between these brain regions and regulated physiological stability through change, especially in relation to the stress response, can be derived from cases in which long-term or chronic stress is applied. A classic example is post-traumatic stress disorder (PTSD).[3] Individuals clinically diagnosed with PTSD have blunted glucocorticoid responses to some types of social stress. PTSD participants in the Trier Social Stress Test fail to show an increase in cortisol concentrations compared to healthy controls, which indicates a failure to mount acute physiological responses to stress events (Wichmann et al., 2017). The lack of change in cortisol concentrations is in part due to marked changes in brain morphology. Patients with PTSD have significantly decreased bilateral grey matter volume in the hippocampus and amygdala (see Karl et al., 2006 for a review). Combat-related trauma experience by soldiers has profound impacts on brain function: significant reductions in the size of the hippocampus and frontal cortex are two well characterized examples (Sussman et al., 2016). These illustrate that allostatic processes are severely disrupted, leading to altered stress responses, even though the homeostatic and programmed rheostatic responses remain intact.

Hierarchical organization of the stress response

Now that we have covered the physiological basis of the stress response and outlined how the amygdala, hippocampus, and frontal cortex are involved in allostasis, we can start to develop a hierarchical organization of physiological stability. Here, the stress response is an excellent example to illustrate how homeostasis, rheostasis, and allostasis form a layered network for coordinating physiological responses to maintain internal stability over short-term timescales. Based on individual differences (e.g., genes), prior experience (e.g., development), or memories, allostatic control of physiological responses, which involves the frontal cortex, amygdala, and hippocampal structures, modifies programmed physiological changes regulated by hypothalamic substrates and homeostatic balance (Figure 8.4). The connections between the frontal cortex, amygdala, hippocampus, and hypothalamus will undoubtedly also include indirect connections through other brain

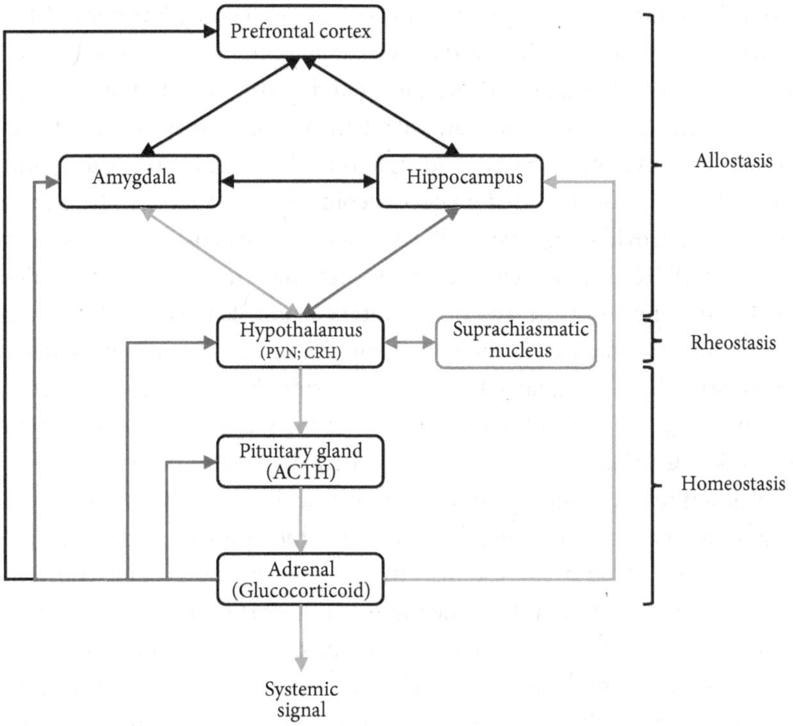

Figure 8.4 Schematic diagram of the neuroanatomical substrates and hormonal control of the stress response.
The hypothalamo-pituitary-adrenal axis provides homeostatic regulation of short-term stability in plasma glucocorticoid concentrations. Input from the suprachiasmatic nucleus in the hypothalamus provides the long-term programmed rheostatic regulated change in circulating glucocorticoid concentrations (e.g., daily rhythm). Cortical brain regions, including the prefrontal cortex, amygdala, and hippocampus shape the stress-induced change in glucocorticoid concentrations that is individual-dependent and based on prior experience. Corticotropin-releasing hormone (CRH) neurons in the paraventricular nucleus (PVN) form the central node for the integration of all three levels of physiological stability.

structures, such as the bed nucleus of the stria terminalis (Radley et al., 2015). These neural inputs provide "top-down" control and modify homeostatic and rheostatic physiological responses to stress event(s).

The homeostatic regulation of glucocorticoid concentrations in the blood provides *primary stasis* that requires a hypothalamic-pituitary-adrenal feedback circuit. The neuropeptide corticotropin releasing hormone (CRH) is produced by neurons in the PVN of the hypothalamus, and these cells project down to the anterior pituitary gland and stimulate the release of adrenocorticotropic hormone. Adrenocorticotropic hormone then travels through the circulatory system and signals to the adrenal cortex to synthesize glucocorticoids. Elevated glucocorticoid concentrations in the circulatory system provide negative feedback whereby glucocorticoids inhibit the activity of PVN CRH which in turn reduces the release of adrenocorticotropic hormone from the anterior pituitary gland. It is the CRH neurons that establish homeostatic set points for the circulating glucocorticoids. Glucocorticoids also signal to the pituitary to stop the release of adrenocorticotropic hormone. An additional pathway via the amygdala provides indirect feedback to CRH neurons to inhibit adrenocorticotropic hormone release.

As described in Chapter 3, many physiological systems are modulated by programmed circadian timing mechanisms, which provide *secondary stasis*. In the context of the stress response, this programmed rheostatic control of CRH and its downstream stress mediator adrenocorticotropic hormone and glucocorticoid synthesis is achieved by two mechanisms. First, the circadian pacemaker, the suprachiasmatic nucleus provides direct input to the PVN and governs the daily rhythmic change in the set point for glucocorticoids. Second, the circadian clock genes in CRH neurons, the cells of the pituitary that produce adrenocorticotropic hormone, and the adrenal gland that produce glucocorticoids provide an intracellular signal to fine tune the functionality of each level of the neuroendocrine axis. This hierarchical control has been evidenced in several studies in rodents that have shown time-of-day differences in the production of corticosterone after a stress event. For example, programmed circadian rheostasis of glucocorticoids in rats results in typically low circulating concentrations of corticosterone (approximately 9 μg/100mg) in the morning, which rise by over 3 times (approximately 31 μg/100mg) in the afternoon (Torrellas et al., 1980). The immobilization test is commonly used to induce a robust glucocorticoid response and illustrate the role of programmed circadian rheostasis. Rats that experience immobilization stress in the morning display a significantly higher elevation in circulating corticosterone, a rise of approximately 40 μg/100mg, while immobilization tests during the afternoon result in an increase of approximately 15 μg/100mg. This phenomenon is also seen in human studies, as cortisol concentrations in humans also peak early in the morning and decline over the course of the day, and increased circulating cortisol concentrations in response to stress events are also entirely dependent on the time of day. This is clearly illustrated with data from the Trier Social Stress Test: the increase in cortisol is 4 times larger when the test is conducted in the morning compared to the evening (Yamanaka et al., 2019), in both men and women participants (Lovallo et al., 2010).

With regard to rheostatic control, there are strong bi-directional interactions between programmed circadian clock mechanisms and circulating glucocorticoids. The impact of disrupted circadian clock gene function has opposing effects on corticosterone levels in mice. If circadian clock genes involved in the positive control of the transcriptional-translational feedback oscillator (i.e., *Bmal1/Clock*) are mutated, the result is a significant reduction in circulating corticosterone, a condition known as hypocortisolism (Leliavski et al., 2014). Mutations to circadian clock genes that control the negative feedback to the oscillator (i.e., Cry) lead to elevated levels of corticosterone, or hypercortisolism (Lamia et al., 2011). Interestingly, mutations to another circadian clock gene implicated in negative feedback, period-2 (*Per2*), results in hypocortisolism and heightened stress-like behavior (Yang et al., 2009).

How do allostatic mechanisms, the *tertiary stasis* level in the hierarchical control of physiological stability, interact with homeostatic and rheostatic control of glucocorticoids? The excitatory inputs from the amygdala, inhibitory inputs from the hippocampus, in combination with the prefrontal cortex, act to either enhance or limit the capacity of the CRH-adrenocorticotropic hormone-glucocorticoid axis to respond to a stress event (Figure 8.4). The capacity to respond to glucocorticoids is determined by the concentration of glucocorticoid receptors, which are widely expressed in the mammalian brain with abundant levels in the frontal cortex and hippocampus (Chao et al., 1989). In rats, the expression of glucocorticoid receptors in the PVN in the hypothalamus has a robust daily rhythm with peak expression levels during the early morning, reaching a nadir shortly before lights off (Herman et al., 1993). This daily change in the expression level of glucocorticoid receptors is not observed in other related brain regions, including the PVN of the hypothalamus, the location of the CRH neurons and rheostatic control. Therefore, changes in glucocorticoid signaling in the hippocampus, and consequential effects on hippocampal function, provide a means for allostatic regulation of the CRH neurons and the primary means of allostatic input to regulate homeostatic set points and regulated rheostatic processes.

As indicated above, one of the main characteristics of allostasis is the role of genetic heritability and individual differences in stress responsiveness driven by the expression of genes, and variation in hippocampal connectivity, which led to differences in the physiological responses to a stress event (Meaney, 2001). This variation can also be driven by experience; for example, the social environment early in development can have a profound effect on the stress response in adulthood. This is exemplified by the role of maternal care in the development of stress responses in rats. Mother-juvenile interactions in the nest consist of maternal-to-offspring directed licking and grooming of the pups, and to facilitate nursing of the pups and increase lactation, a mother will engage in arch-backed nursing (Fleming et al., 1999). The amount of licking and grooming behavior received by a pup in the first week of development impacts the amount of maternal care performed later in life. More "attention" from licking and grooming a female pup receives during development will result in high licking and grooming as mothers. This individual variation in maternal care as an adult is caused by a change in the biochemical structures (i.e., methylation) located upstream of the gene that codes for the glucocorticoid receptor in the pups because of their neonatal care (Weaver et al., 2004). Pups that experience high licking and grooming have low levels of DNA methylation compared to pups that received low maternal attention. Lower DNA methylation in the promoter of the glucocorticoid receptor allows the gene to be transcribed and causes more glucocorticoid receptor to be expressed in the hippocampus. Crucially, the increased receptor expression is persistent, lasts well into adulthood, and results in high licking and grooming behavior to pups. A simple test to establish that the early life experience shapes the individual's glucocorticoid receptor gene methylation levels is to "cross-foster" pups. In this experimental manipulation, pups from mothers that engage in high licking and grooming are reared by mothers that show low maternal behavior, and vice versa (Francis et al., 1999). This manipulation results in pups born to low maternal care mothers showing higher than expected licking and grooming behavior as adults due to the experience early in life. But what is the relationship between the levels of licking and grooming experienced during early life (i.e., development) and physiological responses to stress events? Rats that received low maternal licking and grooming maintain higher levels of circulating glucocorticoids compared to rats that received higher maternal care. Furthermore, the immobilization test produces a much larger induction of glucocorticoid and the heightened level of hormone production persisted longer in rats that received low maternal care (Weaver et al., 2004). These findings highlight that those social environments can induce long-term and specific impacts on the genome (i.e., glucocorticoid receptor) that have significant consequences for how an individual will respond, and mount a physiological response, to a stress event. Data such as these also clearly demonstrate that individual variation in glucocorticoid activity in the hippocampus is the primary mechanism for allostatic control of the stress response and is entirely distinct from rheostatic and homeostatic mechanisms.

Theoretical framework for the hierarchical organization of stress stability

We can now put the three levels of physiological stability together into a single theoretical framework (Figure 8.5). The regulated variable we will focus on is blood glucocorticoid levels that are synthesized and release by the effector, the adrenal cortex. Under homeostatic control the levels of glucocorticoids are maintained at very low blood concentrations. The sensor in which changes in circulating glucocorticoids are identified consists of the glucocorticoid receptors located in the PVN of the hypothalamus. Any change in glucocorticoid levels is then identified by the error detector: the CRH neurons. Deviations in glucocorticoid concentration result in the activation of CRH neurons to initiate negative homeostatic responses to maintain short-term stability. Here the set point is established by the CRH neurons. The anterior pituitary gland forms the controller and responds to CRH by triggering adrenocorticotropic hormone release into the circulatory system. Increased adrenocorticotropic hormone release stimulates more glucocorticoid production by the adrenal cortex, and, conversely, a reduction in adrenocorticotropic hormone release results in lower circulating glucocorticoid concentrations and homeostatic stability is maintained.

Long-term rheostatic control over circulating glucocorticoids is maintained by programmed circadian mechanisms. The set point established by CRH neurons in the PVN receive input from the suprachiasmatic nucleus (SCN). SCN input is directly responsible for the regulated, daily change in circulating glucocorticoid concentrations. Moreover, the transcriptional-translational feedback loop established by circadian clock genes maintains cellular timing mechanisms within each CRH neuron to control time-of-day activity.

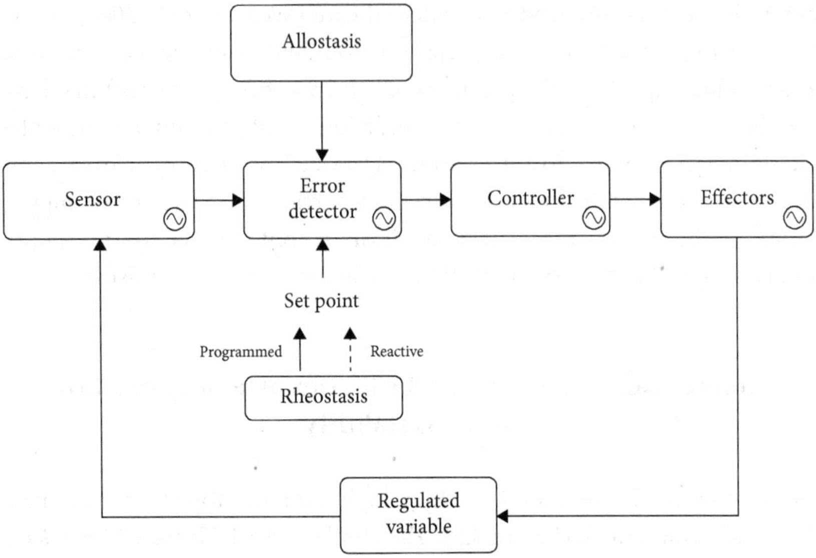

Figure 8.5 Theoretical model for allostasis, rheostasis, and homeostasis. The hierarchical organization of physiological stability builds on the four main components of homeostatic regulation of a variable. The homeostatic set point is maintained within a restricted physiological range and rapidly maintained. Rheostatic mechanisms afford the regulated change in a physiological variable and the set point is maintained by separate genetic, cellular, and neural components. Allostasis provides individual control over a physiological variable and is coordinated by cortical-amygdala-hippocampal circuits that can modify the functional value of the set point (i.e., CRH neurons).

Lastly, allostatic control of glucocorticoid levels is governed by the coordinated activity across the frontal lobe, amygdala, and hippocampus. Glucocorticoid receptor expressing cells in the hippocampus are the predominant input for controlling glucocorticoid concentrations governed by CRH neurons in the PVN in the hypothalamus. In healthy organisms, the input from the hippocampus modifies the homeostatic set point by reducing or elevating the magnitude of circulating glucocorticoid levels and is critical for the termination of the stress event. In pathological conditions, such as allostatic overload, there is increased input from hippocampal cells to the CRH neurons leading to elevated and sustained concentrations of circulating glucocorticoids.

Summary of allostasis and the hierarchical control of physiological stability

This chapter has introduced the concept of allostasis and focused on the control of glucocorticoid concentrations in the blood. The data indicate that complex neural activity is essential for the individual differences in stress responses and place the hippocampus as the central node for allostasis. There are also profound sex differences in what is considered a stress event and in the stress response. Current evidence indicates that the CRH neurons in the PVN are the primary node in which differences in male and female stress responses are controlled (Bangasser and Wiersielis, 2018). A hierarchical organization of physiological stability was proposed to account for the different levels of control over one physiological variable: glucocorticoid concentration in the circulatory system. Allostasis is distinct from homeostasis and rheostasis in that the physiological response involves anticipation of environmental cues that will cause a perturbation in internal stability. The mechanisms of anticipation are not the same as those involved in regulated responses for rheostasis. Anticipation in rheostasis is a programmed event that uses inherited, genetic material. Anticipation in the allostatic control of physiological responses is dependent on prior experience of the same or similar environmental cues and requires an entirely separate set of brain regions. It is important to highlight that allostasis is primarily individual-specific and the concepts outlined in this chapter, unlike homeostasis and rheostasis, display considerable variability in any given population. There are multiple means of shaping physiological stability through early life experiences (i.e., development), inherited genetic variation, and adult life experiences. Damage to the prefrontal cortex, amygdala, or hippocampus can severely disrupt individual stress responses. Disruption in allostatic control, demonstrated by patients with PTSD, highlights that *tertiary stasis*, can be differentiated at neural levels from both *secondary* and *primary stasis* mechanisms. Allostatic mechanisms and the hierarchical organization of stability apply across physiological systems, not just the stress response. Another example could include the hierarchical organization of energy stability and the role of cortical regions in governing glucose, (an)orexigenic hormone, via modulating food intake and appetite. Now, let us turn away from biological explanations that separate physiological stability and discuss how homeostasis, rheostasis, and allostasis are characterized by very different mathematical models.

Notes

1. The structure and function of brain regions varies considerably across the animal kingdom. The key links between limbic (i.e., amygdala, hippocampus) and telencephalon (i.e., cortex; pallium) are homologous across vertebrates. See Butler et al., 2012 for a critical review.
2. Androgen steroids such as testosterone produced by the adrenal cortex are significantly higher in patients with Congenital Adrenal Hyperplasia. Individuals such as Giuseppe Marzo may be genetically female, but high levels of testosterone produced by the adrenal glands will result in an individual that looks like a male later in development. Cases such as these highlight that an individual's sex cannot be defined only by genes or physical appearance.
3. The American Psychiatric Association has developed the Diagnostic and Statistical Manual of Mental Disorders book, abbreviated DSMV (V for fifth version) as a source used to clinically determine whether an individual suffers from PTSD.

9

Modeling physiological dynamics

> Everything around you is mathematics. Everything around you is numbers
>
> Shakuntala Devi (1929–2013)

Shakuntala Devi had a remarkable ability to solve mathematical questions. Born and raised in Bangalore, India, Shakuntala had very little education but could quickly solve highly complex arithmetic problems. Because of her uncommon skill, there was strong interest among other people to witness her abilities, and Shakuntala toured the world solving lengthy mathematical computations. Arguably the most famous event occurred at Imperial College in London, where Shakuntala mentally computed the product of 2 13-digit numbers in only 28 seconds. This achievement was recognized by the Guinness Book of World Records and the accolade of "human computer."

Scientists aspire to devising mathematical formulations that account for observed phenomena. The mathematical equation that most of the public will be familiar with is Albert Einstein's formulation $E = mc^2$; a simple calculation for physical energy (E) where energy is the product of mass (m) and the speed of light (c) squared. The renowned physicist Stephen Hawking (1942–2018) developed an equation that built on the work of Einstein. Hawking radiation is represented as $T=(\hbar c^3)/(8\pi MGk)$ and explains the radiation emitted from black holes. In this equation, T is temperature, ħ is Planck's constant, c stands for the speed of light (from Einstein), 8π indicates the spherical nature of black holes, M is the mass of the black hole, G is Newton's gravity constant, and k stands for Boltzmann's constant for energy of gas particles. Many of the most widely known scientific equations are created by physicists. In this chapter we will discuss a select group of mathematical equations that are derived from dynamical systems theory. The aim is to illustrate that homeostasis, rheostasis, and allostasis are conceptually different from a mathematical perspective and that different equations have been developed for the

On Rheostasis. Tyler John Stevenson, Oxford University Press. © Oxford University Press 2024.
DOI: 10.1093/oso/9780197665572.003.0009

three levels of physiological stability. For each section, the history behind the respective computational approaches is highlighted.

Homeostasis models originated in steam engines and naval ships

The mathematical formula traditionally used to describe the rapid feedback mechanism required for homeostatic stability is based on control theory, which originates in engineering problems and was developed to account for how feedback into a mechanical device is used to influence the function of a working, dynamical system (e.g., engines).

Scottish mathematician and scientist James Clerk Maxwell (1831–1879) is widely recognized for his work on electromagnetic radiation. During his lifetime, engineers needed to know how machines such as steam engines could maintain an exact speed, or steady state (i.e., modern day cruise control). Maxwell demonstrated that feedback control is integral to the function of engines and that a steady state was achieved using a governor. A governor is a device that maintains the velocity of the machine nearly constant, despite changes in the driving power or resistance (Maxwell, 1868). Maxwell's mathematical solution to describe control theory is often compared to the significant advances made by Isaac Newton and Albert Einstein (Tolstoy, 1981). Since then, several additional modifications to control theory have been developed. The Routh-Hurwitz theorem is a mathematical test that made significant contributions to assessment of whether a controlled system is stable; it was independently derived by Adolf Hurwitz (1859–1919) and Edward Routh (1831–1907). More recently, Nicolas Minorksy (1885–1970), a Russian American mathematician and engineer, proposed proportional-integral-derivative (PID) controllers (Minorsky, 1922), which consist of a controlled loop mechanism that uses feedback to continuously monitor control. The PID controller calculates an error value as the difference between a desired set point and a measured process variable and applies a correction that is proportional, integral, and derivative. Concepts of PID controllers developed by Nicolas were fundamental for the installation and testing of automatic steering for the battleship USS *New Mexico* (Bennett, 1984). The fundamental concepts of control theory, such as regulated or dynamical systems, feedback loops, and governors (i.e., controllers) were attractive to early physiologists because of the ease with which the physical models could be applied to physiological processes.

Visual depictions of control theory and dynamical systems often use "block diagrams" to illustrate the associations between each mechanical component. We will be familiar with block diagrams as the design has been applied to the theoretical framework of homeostasis throughout the book (Figure 9.1). Control theory has two types of feedback loops classified as "open-loop" or "closed-loop." In open loop control systems, the regulated variable is independent of the output. A typical example consists of switching on or off an automobile. Closed loop incorporates the concept of feedback control and includes set points (e.g., cruising speed). In closed loop designs, any deviation from the set point is identified and an action initiated to restore the system so that any change is returned to zero (Mayr, 1970). For Cannon, the engineering principles central to closed looped designs relate well to how homeostasis is controlled by classical physiological feedback mechanisms. The simplest equation used in control theory to explain how stability is maintained within a system can be expressed as the measured output of the system (y) (i.e., the regulated variable) given by the sum of the desired state (or set point, *Sp*) and the error (*e*) and can be represented as: $y = Sp + e$. If the error from the desired state is zero, then the equation (and physiological state) is equal to the set point. When the regulated variable is above the set point, the error value will be positive. Conversely, when the regulated variable is below the set point, the error value will be negative.

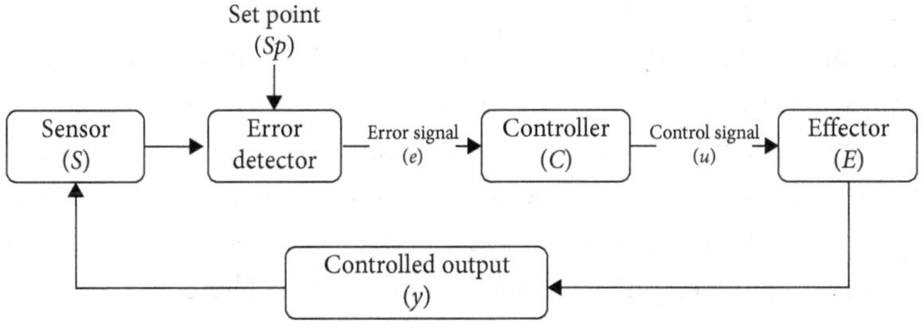

Figure 9.1 A block diagram of a closed-loop transfer function for homeostasis. The regulated variable, controlled output (y) provides feedback to the sensor (S) that is compared with the reference set point (Sp). The controller (C) then takes the error value (e) to change the input value of the control signal (u) to the effector (E). Changes in the output of the effector alter the regulated variable to maintain the desired steady state.

However, dynamic physiological systems require a few more components and the equation can be developed to include a controller (*C*), the system or physiological effector (*E*), and the sensor (*S*) we have discussed in the preceding chapters. We can take the functional values of *C*, *E*, and *S* and plug these values into an equation based on control theory and calculate homeostatic stability as:

$$\textit{Equation 1}: y(s)=\left(\frac{C(s)E(s)}{1+C(s)E(s)S(s)}\right)Sp(s)$$

In this equation, the error value consists of the product of the controller and effector divided by the product of the controller, effector, and sensor. Here, the numerator is the feed-forward gain from the set point or the mechanism to return any deviation in the regulated value to the established set point. The denominator is the closed loop transfer function (i.e., feedback) in which the sensor determines the extent of the deviation. Together, the numerator and denominator form the overall error signal (*e*). Note that homeostatic set points do not vary over time but instead are fixed values.

The model proposed for the homeostatic control of physiological stability is a simple-input-single-output model and many other inputs and functional modifications can be applied. Despite the clear links between homeostatic control systems and control theory formulation, there are limitations to this model. Set point values are generally easy to establish, for example, core body temperature is stable at around 37.5°C. Establishing the appropriate values for the controller, effector, and sensor is more challenging. For example, do the values of sensor use receptor numbers or intracellular signaling pathways? Does the controller use the number or electrochemical activity of warm- or cold-temperature sensitive neurons (or both)? What effectors are used in the equation? Does one select the amount of shivering or some other value that includes the amount of water loss due to sweating. A second limitation is that regulated physiological variables can be altered by one or more input/output components and other "multi-input-multi-output" equations can be modeled. To overcome these limitations, a series of *differential equations* are required to accurately model homeostasis. Differential equations can consist of several computations that can relate one or more known and unknown functional values. Once the differential equations are identified, a transformation called Laplace transform can be performed, which takes real variable values and calculates a complex variable known as *s*-domain (e.g., *C(s)*). This approach has been successfully used to model some forms of homeostasis such as energy stability (Pattaranit and van den Berg, 2008). A full description of the mathematical modeling used in control theory is beyond the scope of this chapter.

Rheostasis models require regulated changes of the fourth dimension: time

Mathematical modeling that incorporates regulated changes across time is generally associated with a different dynamical systems theory and instead uses *time-series* analyses or forecast models. One type of mathematical operation used to describe the functional relations of rheostasis can be adapted from the field of circadian biology. Circadian analyses provide a robust example for how programmed rheostasis is calculated completely differently to homeostasis. At a basic level, rhythmic patterns, whether circadian, menstrual, or seasonal, are predictable oscillations, which are suitable for Fourier analyses. French mathematician Jean-Baptiste Joseph Fourier (1768–1830) demonstrated that any oscillation (or harmonic) could be expressed by the sum of simple trigonometric coefficients and showed how to deconstruct an oscillation into a series of terms that included cosine and/or sine waveforms. This is an extremely useful method that can be used to calculate the period, phase, and amplitude of a waveform.

Fourier analysis was originally applied to circadian datasets by Franz Halberg and colleagues (1965). One of the most common analyses of programmed circadian rheostasis uses time series analyses that include cosinor rhythmometry.[1] In this case, rheostasis is not dependent on feedback control, but instead represents oscillations in the form of a sinewave or cosine wave. In our block designs, one argument was that programmed rheostatic inputs were provided by the circadian clock residing in the suprachiasmatic nucleus and directed to the set point control mechanism (i.e., neuropeptide) for any given regulated variable. For a mathematical model to include both rheostasis and homeostasis, the set point variable must be influenced by a waveform function. To achieve this, the set point will be dependent on a cosinor rhythm. Using Equation 1 above, the "s-domain" would include a differential equation to account for rheostatic changes in the set point. Moreover, programmed rheostatic mechanisms are also dependent on circadian timing mechanisms within the controller, effector, and sensor. The differential equations used for each physiological component (e.g., controller) would also need to include operations that account for rheostatic programs.

Other programmed or reactive rheostatic processes can be incorporated by simply adding additional differential equations into the computation for the set point and not necessarily the error value. For example, the rheostatic contribution to the equation can both include the circadian value using the regression model and add a seasonal programmed rheostatic value.

Information theory and concepts of contingency and constancy have been developed to describe the seasonal functions of neuropeptides in birds and mammals (Stevenson and Ball, 2011). In the case of reactive rheostasis, simple periodicity analyses can be used to model the single bout in which the regulated variable changes. Fourier analysis can be used to construct a periodogram to identify the spectral energy associated with each frequency. Using the human pregnancy example from Chapter 7, the periodogram generated would show a spectral energy peak at around 37–40 weeks.

So far, we have shown that homeostasis is represented best by control theory equations and each component can be adjusted according to rheostasis processes by incorporating cosinor analyses. The next section considers how allostasis can be modeled mathematically and contribute to the hierarchical organization of physiological stability.

Cognitive theory and brain models for allostasis

Modern neuroscience theories have used computer-based systems to describe how different nuclei in the brain function in a network to sense, perceive, and move within an environment, and crucially, how cognitive states govern behavioral decision-making. Computational neuroscience is a specific discipline that develops mathematical models and simulation to understand how the brain controls behavior. Therefore, any mathematical model that we use to account for allostatic control over physiological systems will be based on approaches developed from the field of computational neuroscience.

The field of brain sciences or neuroscience has a strong tradition of using mathematical modeling to understand physiological processes. Two early neurophysiologists Alan Hodgkin (1914–1998) and Andrew Huxley

(1917–2012), working in the Physiological Laboratory at the University of Cambridge, developed a mathematical model using empirical data obtained from their research on the electrochemical properties of the Giant squid axon. Most of their research examined rapid changes in moving particles (e.g., sodium) across cell membranes and how changes in currents permitted communication between and within neurons (Schwiening, 2012). For their pioneering work in understanding the distribution, abundance, and flow of ions across the squid axon, they were awarded the 1963 Nobel Prize in Physiology or Medicine. Arguably, the pinnacle of their collaboration was the publication of a mathematical model of the action potential (Hodgkin and Huxley, 1952), which is quite ingenious and like homeostatic concepts, is based on dynamical systems. Yet, the main difference is that neuronal communication is an open system and does not form a closed-loop structure akin to homeostatic control theory. And despite the similarities of time-dependence with rheostatic concepts, each step in the propagation of the action potential down the neuron axon has a defined space or finite length. Early examples of mathematical formulations such as the Hodgkins-Huxley model were highly informative for attempts to understand how multiple neurons within a group or "network" function together to produce a physiological response.

Donald Hebb (1904–1985) a Canadian psychologist and Professor at McGill University proposed that *neurons that fire together wire together* and established Hebb's Law. In his book *The organization of behavior*, Hebb described how changes in communication between two neurons were responsible for adaptation in brain networks. Hebb's law is a fundamental property of how neurons form networks to produce behavior and regulate physiological processes. One of the first equations to describe the function of neuron networks was conceived by John Hopfield, a professor of molecular biology at the University of Princeton, who created the first mathematical model that described the relationship between two neurons that were active (i.e., produce action potentials) or inactive (i.e., no action potential). The model was easily expanded to include multiple neurons or a neuronal "net" that consisted of binary interactions between a series of neurons (Hopfield, 1982). Based on the activity of any neuron, the sum of the neuronal network could be determined using Hebb's Law. In the model, neurons learn patterns, and exposure to input values form a level of "pattern recognition." The ability to mathematically model neuron networks and the association between neurons has provided the basis for memory research for decades. At the simplest level, Hopfield networks describe the relation between two neurons or "nodes" and Hebb's law that neurons wired together, fire together.[2]

Early applications of dynamical systems to develop models of brain activity were described as behavioral dynamics and focused on the bidirectional relations of an individual's behavior and the individual's environment. Modern versions include how the individual's behavior changes in relation to the environment and is referred to as an open dynamical system. This approach allows for the investigation of the cognitive and/or behavioral processes of an individual within the environment, but also the same individuals cognitive and behavioral state in the absence of the environment. When taking the internal physiological state of the individual into consideration, models must include aspects of the "body system." One approach that attempts to model cognition and behavioral outputs of an organism, while considering the role of bidirectional relations between an organism's behavior and interactions with the environment, is embodied cognition theory. Embodied cognition is based on a mathematical model of dynamical systems and is well suited to account for allostasis. As allostatic responses are dependent on perception (e.g., potential stress event) and the physical state of the individual (e.g., age, fitness), the use of alternative dynamical systems such as embodied cognition is appropriate.

Using the example of the stress response, the mathematical input of allostasis for the regulation of physiological stability is represented as the input from the hippocampal neural network into the set point maintained by the paraventricular neurons. In this approach, we can apply Hopfield network models to allostatic input from hippocampal neurons to neural networks that control set points (e.g., neuropeptides). One clear example is the glucocorticoid receptor expressing neurons in the hippocampus that project to CRH neurons in the PVN. The set point value is adjusted based on Hopfield network inputs from the hippocampus (or other cortico-limbic nodes). Hopfield networks have now expanded to include measurements of "energy," which determines the stability of discrete connections. When stability is maintained, whenever the state of any "neuron" changes, the energy function will decrease. Most neural connections now are represented as continuous Hopfield networks and include time as the continuous variable. However, the next challenge is to incorporate differential equations to model allostatic regulation of set points outlined in Equation 1.

Conclusion and limitations of the mathematical models

In this chapter we have discussed three very different conceptual frameworks that describe homeostasis, rheostasis, and allostasis. Homeostasis has traditionally been modeled using closed-loop transfer functions based on control theory; rheostasis uses a range of different models that include time-dependent components; and allostasis can use other forms of dynamical systems. Developing a single unifying model to describe the hierarchical organization of physiological stability will undoubtedly be a tricky pursuit and differential equations that include multiple computations that incorporate feedback, time-series, and Hopfield networks will be necessary. One major limitation to developing accurate models of physiological stability is the significant lack of measured values for any regulated variable that were collected based on rheostatic and allostatic experimentation. For example, most studies do not examine regulated variables or other physiological components (i.e., controller/effector) across circadian phases. Studies that investigate the impact of allostasis or allostatic load rarely take into consideration programmed rheostatic input. Overcoming these types of limitations requires massive sample collection from individuals. The analyses would have to consider not only the values associated with the regulated variable (e.g., circulating cortisol) but also the input and output values of the sensor, controller, and effector. And finally, the required processing power to compute the equation would likely demand a quantum computer for the vast number of physiological values and complex computations.

Notes

1. Cosinor rhythmometry is calculated as: $y(t) = M + A\cos\left(\frac{2\pi t}{\tau} + \phi\right) + e$ (Refinetti et al., 2007). M is the MESOR (midline statistic of rhythms, or the mean amplitude), A is the amplitude (a measure of half variation in the cycle), ϕ is the acrophase (the time of overall high values recurring in each cycle), τ is the period (duration of one cycle), and e is the error term.
2. Hopfield networks are computed as: $w_{ij} = \frac{1}{N} p_i p_j$. Here, w_{ij} is the strength of the relation between neurons i and j, N is the number of nodes in the network, and p_i is the value of the i-th node.

10
Challenges to physiological anticipation

> I became insane, with long intervals of horrible sanity.
>
> Edgar Allen Poe (1809–1849),
> letter to George Washington Eveleth, 1846

Just after the *period* gene had been identified, sequenced, and shown to be critical for establishing daily rhythms in behavior (Hardin et al., 1990), Nicholas Mrosovksy concluded his book *On rheostasis* with a brief note on the potential of molecular biology to make substantial gains into understanding the mechanisms and systems involved in the physiological control of regulated variables. Now, over 20 genes are known to regulate the circadian transcriptional-translational feedback loop that is central for programmed circadian rheostasis. Whole genome sequencing has now become much easier, and the application of complex genome-wide association analyses has allowed the identification of novel genomic regions involved in programmed and reactive rheostasis, as well as allostasis. Our understanding of the control of physiological processes has also been advanced by information gained from studies of the effects of simple gene modifications and alterations to neuronal circuits. This book has brought together approximately 30 years of the advances achieved with molecular biology, with the well-described physiological, lesion, and hormonal manipulations Nicholas used to characterize rheostasis, to summarize our understanding of the control of physiological processes. A couple of questions remain, however: (1) What are the consequences of long-term disruption of rheostatic processes to animal health and well-being? (2) How do regulated variables impact programmed or reactive responses in another regulated variable? In this chapter we will address the first question by describing disrupted rheostatic processes and will use shift work in humans and climate change as current examples of detrimental effects on regulated changes in physiological stability. To address the second question, I will describe Nicholas Mrosovsky's concept of *second order rheostasis*, defined as the modulation of rheostasis by altering the set point of the regulatory system. Examples of altered set points caused

On Rheostasis. Tyler John Stevenson, Oxford University Press. © Oxford University Press 2024.
DOI: 10.1093/oso/9780197665572.003.0010

by conflicts between regulated variables will be used to illustrate the impact of changes in one regulated variable on the set point of another regulated variable. We will finish with a discussion of the potential consequences of adapting to the final frontier and the impact of space exploration on the regulation of physiological processes in humans and domesticated organisms.

Disrupted programs and reactions

The importance of physiological stability is highlighted by the effects of disruption to the anticipated "normal" changes in physiological variables, which can lead to pathological conditions. We have previously discussed how some of the most significant disruptions to the rheostatic and allostatic control systems that we are exposed to occur because of our own actions (i.e., anthropomorphic) and the effects of societal or climate change. The best characterized or most pronounced changes are those that have occurred over the past 200 years as a consequence of the industrial revolution: the associated lifestyle changes have major impacts on the health and wellness of humans and animals. We live in a near 24/7 society in which our activity patterns include shift work, travel, and social media which continually disrupt the stability of evolutionary ancient, programmed changes within our physiological systems. Figure 10.1 illustrates the principles of how programmed and reactive rheostasis is disrupted by modern life. One specific example is the dampening of the programmed rheostatic changes in melatonin secretion by the increased duration of light exposure because of artificial light (red line, left graph), the consequences of which are negative effects on mental health and cardiovascular and metabolic function. Mice exposed to increased dim light are heavier, due to increased feeding during the "light phase" when metabolism is low and glucose tolerance reduced (Fonken et al., 2010). The circadian rhythm in melatonin can also be shifted (blue line, left graph) because of jetlag and shift work. For example, nurses who experience rotating shift work schedules are more susceptible to becoming obese and/or experiencing metabolic syndrome and type-2 diabetes (Pan et al., 2011).

Circadian misalignments can also lead to other physiological changes that underlie emotional, cognitive, and bodily (i.e., somatic) responses, such as changes in immune function and an increased risk of cancer (reviewed in Wulff et al., 2010). Emotional responses include mood fluctuations, increased irritability, frustration, impulsivity, decreased motor performance,

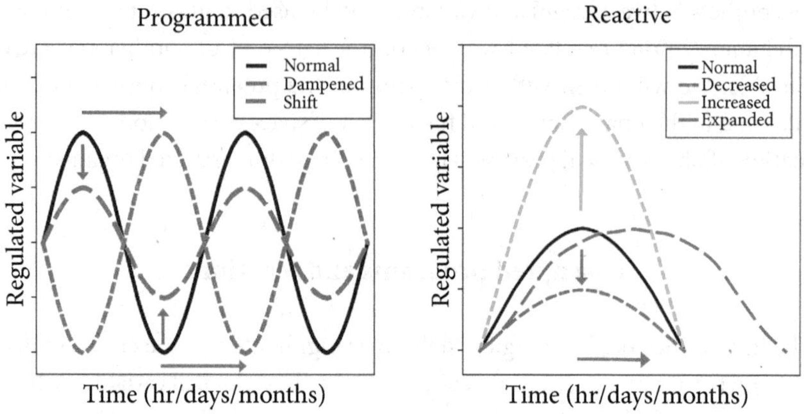

Figure 10.1 Examples of disrupted programmed and reactive rheostasis
Typical programmed (left) and reactive (right) rheostatic process are indicated by the black line. A disruption in the programmed regulated changes in set points can occur by a decrease in the oscillation amplitude (red arrows) and is illustrated by a dampening in the waveform (red line). Other types of disruption to programmed changes in the set point value include a shift in the waveform (blue arrow and lines). Similar types of disruption can occur in the reactive rheostatic control of set points indicated by a decrease (red) or increase (green) in the amplitude of the regulated variable. Finally, the shape of the reactive rheostatic response can be expanded to occur over a longer timescale indicated by the blue line.

and mania. Changes in cognitive responses often involve reduced ability to sustain attention and alertness, decreased memory performance, and reduced multitasking, decision-making, and creativity. Somatic responses are generally characterized by tiredness; chills and changes to immune function can lead to increased susceptibly to viral infections. The importance of centrally programmed circadian rheostasis is also shown by examination of psychiatric disorders and neurological diseases, most of which are associated with marked changes in programmed circadian function. For example, increased sleep disturbances are commonly reported in patients that suffer from Alzheimer's disease and schizophrenia. Schizophrenia is also often associated with arrhythmic activity patterns with individuals showing increased activity across a 24-hour period.

The climate change crisis has received much attention for its effects on global temperatures. However it also poses a threat to physiological stability, as it is causing long-term shifts in the patterns of annual rhythms in environmental cues. In so doing it is leading to disruptions to programmed rheostasis of physiological processes with timescales longer than a day (i.e., seasonal rhythms). The Intergovernmental Panel on Climate Change (IPCC) uses the pre-industrial global surface temperature from 1850 to 1900 as the reference period for monitoring changes in current and future climates. Over the past 20 years, we have experienced an increase of 1.09°C in global surface temperatures. The IPCC predicts the 1.5°C limit will be exceeded by 2040 and will result in more desertification of land area, decreased precipitation, loss of biodiversity, the melting of glaciers, more ocean acidification, and a sea level rise (IPCC, 2022). Local changes in surface temperatures, however, can already be seen to be having a dramatic impact on programmed circannual rheostasis across multiple trophic levels. In the United Kingdom, the past 50 years has witnessed a gradual increase in average temperatures from January to August. Higher temperatures in April and May have resulted in the earlier emergence of leaves and earlier flowering and fruiting in the spring. Butterflies and moths are also taking to their first flights earlier in the year. Birds are arriving from migration sooner and laying eggs earlier than historical records indicate (Thackeray et al., 2010). However, not all plants and animals shift their phenology (i.e., response to cyclical or seasonal timing) at the same rate, which can result in a "mismatch" between life history stages for some interdependent organisms. For example, the shift in insect larvae emergence has advanced at a faster rate than egg laying in some bird species. This mismatch can be detrimental as the insect larvae used by many birds to feed their offspring are no longer available when they hatch.

Who shall judge the judges?

Systems that maintain regulated variables around a set point can also be adjusted (or disrupted) by other, additional internal mechanisms. Here *second order rheostasis* is proposed to provide a superimposed control mechanism that can modulate the rheostatic control over one or more regulated physiological variables. There are several ways in which rheostatic regulation of one variable can be modulated by another regulated variable. Figure 10.2 has four examples that serve to provide a general overview of the types of interactions between different rheostatic control mechanisms. First, there can be conflict between two programmed rheostatic processes in which

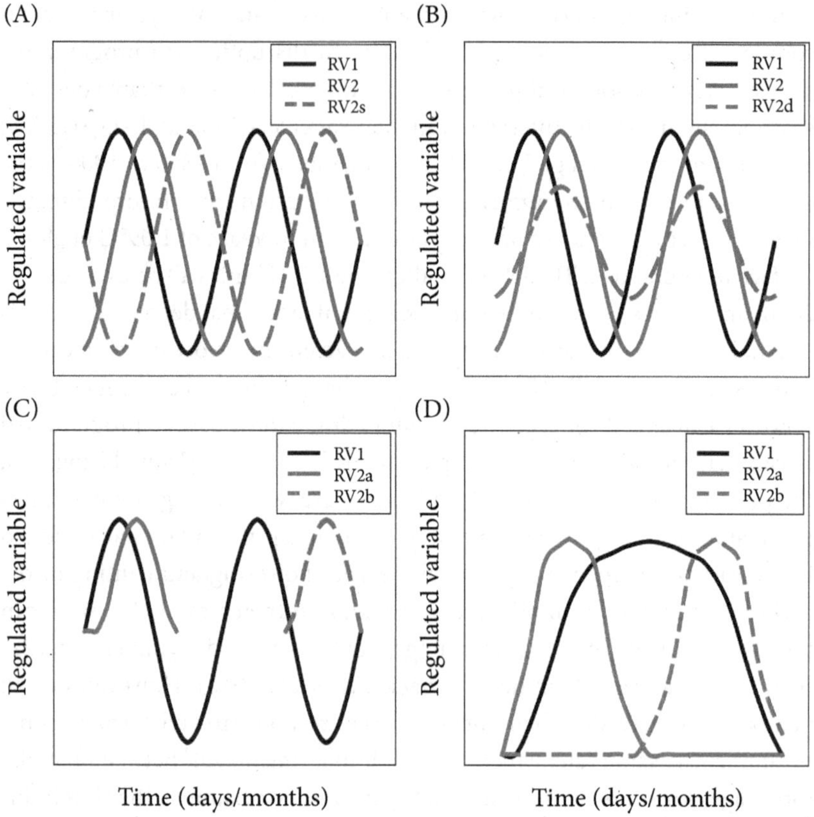

Figure 10.2 Conflicts between programmed and reactive rheostasis. Representative examples of conflict between set points of regulated variables (RV). (A) represents conditions in which there is a shift in the waveform of a second regulated variable (RV2s) controlled by programmed rheostasis. (B) Under some conditions, the amplitude of the programmed rheostatic control of a second regulated variable can be decreased (RV2d). Conflict can also develop between programmed and reactive rheostatic processes. (C) A regulated variable driven by reactive rheostasis can modulate the programmed rheostatic control of another regulated variable if the physiological change occurs at one phase (RV2a) compared to an anti-phase event (RV2b). Lastly, set points can also be modified during conflict between reactive rheostatic processes. (D) The set point for one reactive rheostatic process can be modified if another reactive rheostatic processes develops early (RV2a) or later (RV2b) during different phases in the regulated physiological response of RV1.

a regulated variable has a shift in the set point (Figure 10.2A). Two related variables show a regulated daily waveform change in the set point, variable 2 (RV1) being slightly delayed (RV2) relative to variable 1. However, problems arise when the programmed rheostatic control of a regulated variable is shifted, in this case variable 2 (RV2s), because of, for example, jetlag or shift work. This results in a change in the relationship between the two regulated variables, which can result in physiological disturbance and impaired or loss of function. In this regard, a plethora of research has shown that jet-leg induced changes in physiology can lead to a desynchrony in the signaling across programmed circadian rheostatic processes. The second example (Figure 10.2B) illustrates a dampening rather than a shift of the programmed circadian rheostatic process in regulated variable 2. In this case, the amplitude change in the regulated variable is reduced, but in extreme conditions all rhythmicity could be lost.

Previous attempts to describe second-order rheostasis focused on lesions to hypothalamic nuclei and the impact on regulated variables, such as growth rates, fat deposition, and exaggerated diurnal rhythms in core body temperature (Mrosovsky, 1990). Advances in targeted genetic mutations in mouse models has permitted a greater understanding of the molecular and cellular basis of second order rheostasis. For example, the impact of disrupted programmed circadian rheostasis in one tissue and regulated variable can have significant impact on programmed circadian rheostasis in a second tissue and associated regulated variable. Cutting edge genomic modifications make it possible to ablate the circadian clock *Bmal1* gene in mice in a tissue-specific manner. Greco and colleagues (2021) created a mouse with a mutation in *Bmal1* that was only present in muscle tissue. The result was severe dysregulation in daily clock function in muscle tissue, as would be expected. However, the impact was not limited to muscle tissues, as there were reduced amplitudes in circadian transcript expression in the liver and a significant shift in the peak phase of oscillating genes. Such evidence supports the conjecture that the control of one rheostatic system (i.e., muscle) can influence another rheostatic system (i.e., liver) to modulate the way in which physiological stability is maintained.

An excellent example that illustrates how programmed and reactive rheostatic processes interact and modulate physiological stability is the time of day at which an individual is infected with a pathogen (Figure 10.2C). Healthy Siberian hamsters exhibit daily rhythms in core body temperature, with a temperature of approximately 37.5°C during the dark phase, which declines to 36.4°C during the light phase. An injection of the gram-negative bacteria lipopolysaccharide during the light phase will increase the set point a full degree Celsius and induce a fever that lasts for ~14 hours. However, if the lipopolysaccharide injection is administered shortly before lights off, the set point is only raised to approximately 38°C, and core body temperature returns to normal shortly after the dark phase. Interestingly, daily rhythms in core body temperature are dampened on the subsequent daily cycle. These data highlight that reactive fever rheostatic processes have marked differences in the development and maintenance of set point regulation for core body temperature that are dependent on when in the programmed circadian rheostatic cycle the infection occurs. If hamsters are rendered completely arrhythmic and do not possess a functional programmed circadian rheostatic regulation of core body temperature, core body temperature is maintained around 36.3°C. In these individuals if lipopolysaccharide is administered either shortly before lights off or just after lights on, the magnitude and duration of the increase in core body temperature is nearly identical (Prendergast et al., 2015). Similar findings are reported in mouse models and deletions in the circadian clock gene, *Bmal1*, illustrate that several immune responses are significantly enhanced in rodents with disrupted programmed circadian changes in rheostasis (Edgar et al., 2016).

Conflict between two reactive rheostatic processes has markedly different outcomes that are dependent on the phase in which one response develops (Figure 10.2D). Using the time of infection again, the change in a regulated variable 1 over the course of the reactive rheostatic response will be different if the regulated variable 2 occurs early or late over the period of regulated variable 1. For example, malaria infection early in gestation results in greater likelihood of a pre-term birth compared to infections later in gestation (Elphinstone et al., 2019). The increase in pre-term birth may be associated with disruption to the reactive rheostatic processes involved in angiogenesis and metabolism during the 24 weeks into gestation. Altogether, these few examples illustrate that regulated set points and associated outcomes can be modulated by conflict between rheostatic processes. Unfortunately, our knowledge of the molecular and cellular basis of disruption to reactive rheostasis is limited. We are now at a time in which studies can incorporate advanced genomic manipulations, big-data analyses, or complex meta-analyses to improve our understanding of how set points are regulated and modulated by second order rheostasis. Only with an improved understanding of the mechanisms by which rheostasis functions and differs compared to homeostasis and allostasis can basic research and medical sciences improve the physical, mental, and cognitive well-being of humans and other animals.

Physiological stability beyond Earth

The twenty-first century holds the potential to witness major and exciting changes for space exploration. We now have the capability for extraplanetary travel and the realistic opportunity for continued human presence on the Moon and Mars. This will come with its own challenges regarding the physiological stability of astronauts and extraplanetary communities, as humans have evolved under and are adapted for the conditions present on Earth. A substantial amount of NASA research into astronauts' health and well-being has focused primarily on daily rhythms in the sleep-wake-work schedule. Professor Charles Czeisler is the Baldino Professor of Sleep Medicine and Director of the Division of Sleep Medicine at the Harvard Medical School and also team leader of the Human Factors and Performance Team at the NASA National Space Biomedical Research Institute. For over 40 years Czeisler has studied the human circadian clock and the role of melatonin and has recently developed a schedule to facilitate circadian adaptation in astronauts to improve sleep, alertness, and performance while working on the International Space Station (ISS). On the ISS, despite approximately 15 sunrises and sunsets during each 24-hour period, astronauts engage in scheduled periods of sleep (~8.5 hours), work (~6.5 hours), exercise (~2 hours), and eating (~2 hours). The overall schedule mimics, and is maintained close to, the Earth's 24-hour schedule, so that the astronaut's physiological stability can be protected. Several steps can be implemented as countermeasure to some major changes in the environment that can impact an astronaut's physiological stability, such as the absence of gravity. Exposure to anti-gravity environments can cause severe morphological changes in humans, such as reduced muscle mass, bone loss, and changes in sensorimotor skills. Radiation exposure causes physiological changes, including shifts in body temperature, reduced immune function, and increased risk of cancer, degenerative diseases, and radiation sickness. Astronauts can anticipate these new gravity and radiation environmental cues and can minimize detrimental impacts on physiological stability using radiation shielding, light technologies, exercise, pressure devices, thermal control systems, and enhanced medical monitoring. Living on the Moon or Mars for prolonged periods will come with several additional complications that will also challenge homeostatic stability. The moon is a "sterile" environment, meaning it lacks geophysical variation. The moon does not rotate. Instead, the lunar day, or the time it takes the sun's light to return to the same place on the moon, lasts for approximately 29.5 days. This period consists of a little over 2 weeks of constant daylight and then 2 weeks of night: terrible for daily light-dark rhythms, but excellent for lunar rhythms. The Moon also cannot support an atmosphere, which precludes long term, seasonally rhythmic information that can be derived from changes in precipitation, temperature, or even gas movements (e.g., wind). How we will overcome the challenges this relatively static celestial environment will pose to our bodies and our health, or that of any other living organism, is currently unknown. In the "medium to long term" (i.e., decades to centuries), for successful human habitation, all structures established on the lunar surface will need to be capable of mimicking the daily and annual rhythms in light schedules and air movements that would be experienced on earth. This will be required so that rheostatic systems in humans and other life forms we take with us will experience little disruption beyond impacts associated with radiation and changes in gravity.

How will humans, and domesticated plants and animals, cope with living on another planet that does have geophysical rhythms but which are different from those of Earth? On Mars, there is a daily photoperiod cycle that is very similar to Earth, so in that respect it may be easier for us to adapt to the Martian environment. The Martian solar day, called "sol," lasts 24 hours, 39 minutes, and 35 seconds. The 40-minute expansion in the solar day may not have severe detrimental impacts on human colonizers, as the intrinsic period of the programmed circadian rheostasis in human averages 24 hours and 11 minutes (Czeisler et al., 1999). Nagin Cox, a graduate of the Air Force Institute of Technology and now the Tactical Mission Lead for the Mars Curiosity Rover and Perseverance Rover lives in both Earth and Martian time and is intimately aware of the demands of living in both.[1] However, seasonal programs will be significantly disrupted as Mars takes 687 days to orbit the sun, so the Martian year is almost twice as long as an Earth year. Like Earth, Mars has an axial tilt of approximately 25°, which results in four Martian seasons, but they will not align with any earthly seasonally anticipated changes in physiological function for humans and other life forms living on Mars. Adapting to a new annual Martian period would be expected to impact programmed circannual rheostasis in immune function, endocrine systems, and mental health. This could include seasonal changes in the number and types of immune cells (Dopico et al., 2015), endocrine function (Tendler et al., 2021), and even brain responses to cognitive tasks (Meyer et al., 2016), which could impact health and longevity. There will also be similar mismatches in short- and long-term timing of physiological function on the other planets in our solar system, and health and well-being impacts, as they all have significantly faster rotations and consequently reduced solar day lengths. The solar day length for Uranus is approximately 17 hours 14 minutes, Neptune, 16 hours 6 minutes, Saturn, 10 hours 34 minutes, and Jupiter 9 hours 56 minutes. Several fascinating hypotheses can be derived to predict how programmed and reactive rheostatic processes will change after prolonged exposure to the Martian year, as well as the potential for adaptation across generations.

Advancing our understanding of rheostasis

We have provided clear evidence for homeostatic, rheostatic, and allostatic control of physiological stability. Evidence at multiple levels of biological organization (e.g., genetic, neural) along with mathematical models show that different physiological processes exist and function together to maintain the fitness of organisms. Increased awareness and understanding of the hierarchical organization of physiological stability will improve how we tackle several issues, such as patient-centered approaches to medical care, domestic animal welfare, and strategies to mitigate or adapt to climate change. The ability to manipulate genetic control systems mean that mouse models afford exciting opportunities to identify how one controller or effector tissue can have profound effects on the regulated changes in the set points for other endocrine systems. The explosion in nucleic acid sequencing, proteomics, and "connectomes" has unlocked a range of genes, proteins, cells, and circuits that are involved in homeostatic balance and provides us with opportunities to establish how sensors, controllers, and effectors function under different set point values. Using advanced brain imaging and whole-body imaging approaches we not only now have the ability for real-time dissociation of neural pathways and study of physiological responses under healthy conditions, we also can investigate the impact of disruptions in pathological conditions. This will allow us to gain a greater understanding of how most neurological conditions and degenerative diseases are due to long-term changes in regulated variables, or an inability to maintain stability through change. As we attempt to move the field forward, one of the biggest limitations is the lack of "big data." The development of accurate models of rheostasis will require monitoring several physiological processes simultaneously over multiple rheostatic cycles. Large datasets, such as the UK BioBank, are excellent resources that will allow us to create new hypotheses using the plethora of data available. The astute researcher will need to ensure a clear scientific approach that incorporates multiple working hypotheses (Chamberlin, 1890) and strong inference (Platt, 1964). The papers by Chamberlin and Platt should be the foundation from which advances in rheostasis are spearheaded to permit rapid and solid scientific progress.

Note

1. Nagin Cox has a very nice TEDx talk (https://youtube/9qd36_QPxSY) on the logistics of living in Earth time (i.e., days) and working in Martian time (i.e., sols). The extra 40 minutes in sol time caused a gradual change in daily work-life schedules resulting in phases of nightshifts and dayshifts.

Glossary

Allostasis—the physiological process to maintain stability through a temporary change in homeostatic set point

Circadian—an endogenously generated oscillation in a physiological variable that maintains a period approximating a 24-hour cycle

Circannual—an endogenously generated oscillation in a physiological variable that maintains a period approximating a 24-month cycle

Control Theory—a field of mathematics that uses engineering and mechanical principles to describe dynamical systems

Controller—cells or neuropeptides that govern the synthesis of a regulated variable by an effector

Cycle—a series of physiological changes that occur in sequential order and repeat with regularly spaced periods

Daily—a regular oscillation that occurs over a 24-hour period and is entrained in part by external cues such as light

Detector—a component of homeostatic systems that binds to and monitors the level of a physiological variable

Effector—a cell or tissue that produces and secretes a physiological variable

Endogenous—a genetic or physiological process that originates from within an organism

Entrainment—the alignment or synchronization with the external environment of endogenously generated genetic or physiological processes

Fourier Analysis—a mathematical operation taking general functions and decomposing them into simpler trigonometric or exponential functions with definite frequencies

Homeostasis—an independent regulatory system by which an organism maintains physiological stability, achieved within a dynamical system in which tightly regulated deviations around an established set point are rapidly corrected

Hopfield Networks—a mathematical model that includes recurrent neuronal connections and calculates associations to approximate learning and memory in a network

Hormone—a chemical messenger that travels in the circulatory system and acts on cells and tissues to induce a physiological change

Infradian—regular oscillations that occur with a period that is longer than 24 hours

Innate—a physiological or behavioral process that is determined by genetically inherited information

Neuropeptide—a protein derived from gene sequences and synthesized in neurons in the brain

Oscillation—a regular, repetitive variation in movement around a central point that is characterized by a defined period or length

Phase—the point of an oscillation at a defined location such as the peak or trough of the waveform

Photoperiod—the duration of light and darkness experienced over a 24-hour period

Poikilostasis—a model of physiological stability developed by Wayne Kuenzel to account for changes in the autonomic nervous system of migrating birds

Programmed rheostasis—repeated and regular regulated changes in the set point of a physiological variable

Reactive rheostasis—a physiological response that is characterized by a regulated change in the homeostatic set point for a defined period

Reactive scope model—a physiological model generally applied to stress responses that incorporates both predictive and reactive ranges of regulated variables; deviations exceeding the normal range lead to homeostatic overload or failure

Regulated variable—output signal of a homeostatic effector that is maintained within a narrow and constant range of values

Rhythm—a regular and repeated sinewave in a regulated physiological variable

Seasonal—a regular oscillation that occurs over a 12-month period and is entrained in part by external cues such a light

Set point—a physiological value around which a desired steady state is maintained to establish stability

Ultradian—regular and repeated oscillations that occur during a 24-hour interval; usually consists of periods of 4–6 hours

Zugunruhe—a German word describing the night-time restlessness that occurs during the migratory phase in birds

References

Ali S, Balachandran K, O'Malley B. 2020. Ninety years of progesterone: the 'other' ovarian hormone. J. Mol. Endocrinol. 65:E1–E4.

Allen AP, Kennedy PJ, Cryan JF, Dinan TG, Clarke G. 2014. Biological and psychological markers of stress in humans: focus on the Trier Social Stress Test. Neurosci Biobehav. Rev. 38:94–124.

Amoss M, Burgus R, Blackwell R, Vale W, Fellows R, Guillemin R. 1971. Purification, amino acid composition and N-terminus of the hypothalamic luteinizing hormone releasing factor (LRF) of ovine origin. Biochem. Biophys. Res. Commun. 44:205–210.

Anderson DM, Keafer BA. 1987. An endogenous annual clock in the toxic marine dinoflagellate *Gonyaulax tamarense*. Nature 325:616–617.

Anthony RV, Limesand SW, Fanning MD, Liang R. 1998. Placental lactogen and growth hormone. In Bazer FW. (Ed.) *Endocrinology of pregnancy* (pp. 461–490). Totowa, NJ: Humana Press.

Aponte Y, Atasoy D, Sternson SM. 2011. AGRP neurons are sufficient to orchestrate feeding behaviour rapidly and without training. Nat. Neurosci. 14:351–355.

Archer ZA, Findlay PA, McMillen SR, Rhind SM, Adam CL. 2004. Effects of nutritional status and gonadal steroids on expression of appetite-regulatory genes in the hypothalamic arcuate nucleus of sheep. J. Endocrinol. 182:409–419.

Aschheim S, Zondek B. 1928. The diagnosis of pregnancy: from the urine by demonstration of the hormone of the pituitary body. Klinische Wochenschrift 7:1404–1411.

Aschoff J. 1955. Tagesperiodic bei Mausestammen unter konstanten Umgebungsbedingungen. Pflugers Arch. 262:51–59.

Aschoff J. 1983. Circadian control of body temperature. J. Therm. Biol. 143:143–147.

Baccala R, Gonzalez-Quintial R, Lawson BR, Stern ME, Kono DH, Beutler B, Theofilopoulos AN. 2009. Sensors of the innate immune system: their mode of action. Nature Rev. Rheumatol. 5:448–456.

Bailey M, Silver R. 2014. Sex difference in circadian timing systems: implications for disease. Front. Neuroendocrinol. 35:111–139.

Bakker KM, Martinez-Bakker ME, Helm B, Stevenson TJ. 2016. Digital epidemiology reveals global childhood disease seasonality and the effects of immunization. Proc. Natl. Acad. Sci. 113:6689–6694.

Ball GF, Silver R. 1983. Timing of incubation bouts by ring doves (*Streptopelia risorii*). J Comp. Psychol. 97:213–225.

Bangasser DA, Wiersielis KR. 2018. Sex differences in stress responses: a critical role for corticotropin-releasing factor. Hormones 17:5–13.

Bao AM, Liu RY, van Someren EJW, Hofman MA, Cao YX, Zhou JN. 2003. Diurnal rhythm of free estradiol during the menstrual cycle. Eur. J. Endocrinol. 148:227–232.

Bao R, Onishi KG, Tolla E, Ebling FJP, Lewis JE, Barrett PJ, Prendergast BJ, Stevenson TJ. 2019. Genome sequencing and transcriptome analyses of the Siberian hamster hypothalamus identify mechanisms for seasonal energy balance. Proceedings of the National Academy of Sciences 116:13116–13121.

Barbour LA, Shao J, Qiao L, Leitner W, Anderson M, Friedman JE, Draznin B. 2004. Human placental growth hormone increases expression of the P85 regulatory unit of phosphatidylinositol 3-kinase and triggers severe insulin resistance in skeletal muscle. Endocrinol. 145:1144–1150.

Barrett P, Ebling FJP, Schuhler S, Wilson D, Ross AW, Warner A, Jethwa PJ, Boelen A, Visser TJ, Ozanne DM, Archer ZA, Mercer JG, Morgan PJ. 2007. Hypothalamic thyroid hormone catabolism acts as a gatekeeper for the seasonal control of body weight and reproduction. Endocrinol. 148:3608–3617.

Bartness TJ, McGriff WR, Maharaj MP. 1991. Effects of diabetes and insulin on photoperiodic responses in Siberian hamsters. Physiol. Behav. 49:613–620.

Baylies M, Bargieelo T, Jackson R, Young M. 1986. Changes in the abundance or structure of the *per* gene product can affect the periodicity of the Drosophila clock. Nature 326:390–392.

Beck B. 2006. Neuropeptide Y in normal eating and in genetic and dietary-induced obesity. Philos. Trans. R. Soc. Lond. B Biol. Sci. 361:1159–1185.

Belchetz PE, Plant TM, Nakai Y, Keogh EJ, Knobil E. 1978. Hypophysial responses to continuous and intermittent delivery of hypothalamic gonadotropin-releasing hormone. Science 202:631–633.

Bennett KD. 1990 Milankovitch cycles and their effects on species in ecological and evolutionary time. Paleobiol. 16:11–21.

Bennett S. 1984. Nicholas Minorsky and the automatic steering of ships. Control Systems Magazine IEEE 4:101–105.Björkström NK, Ljunggren HG, Michaëlsson J. 2016. Emerging insights into natural killer cells in human peripheral tissues. Nat. Rev. Immunol. 16:310–320.

Blackburn DG. 2015. Evolution of vertebrate viviparity and specializations for fetal nutrition: a quantitative and qualitative analysis. J. Morphol. 276:961–990.

Bliss M. 1992. Banting: a biography. Toronto: University of Toronto Press.

Blomqvist A, Engblom D. 2018. Neural mechanisms of inflammation-induced fever. Neuroscientist 24:381–399.

Bolborea M, Dale N. 2013. Hypothalamic tanycytes: potential roles in the control of feeding and energy balance. Trends Neurosci. 36:91–100.

Bolborea M, Pollatzek E, Benford H, Sotelo-Hitschfeld T, Dale N. 2020. Hypothalamic tanycytes generate acute hyperphagia through activation of the arcuate neuronal network. Proc. Natl. Acad. Sci. 117:14473–14481.

Borden EC, Sen GC, Uze G, Silverman RH, Ransohoff RM, Foster GR, Stark GR. 2007. Interferons at age 50: past, current and future impact on biomedicine. Nat. Rev. Drug Discov. 6:975–990.

Brelje TC, Scharp DW, Lacy PE, Ogren L, Talamantes F, Robertson M, Friesen HG, Sorenson RL. 1993. Effect of homologous placental lactogens, prolactins, and growth hormone on islet B-cell division and insulin secretion in rat, mouse and human islets: implication for placental lactogen regulation of islet function during pregnancy. Endocrinol. 132:879–887.

Bremner WJ, Vitiello MV, Prinz PN. 1983. Loss of circadian rhythmicity in blood testosterone levels with aging in normal men. J. Clin. Endo. Metab. 56:1278–1281.

Brown-Sequard CE. 1856. Recherches experimentales sur la physiologie el la pathologie des capsules surrenales Arch. Generales de Med. 5:385–401.

Buckley SJ. 2003. Lotus birth: a ritual for our times. Midwifery Today 67:36–38.

Buntin JD. 1986. Role of prolactin in avian incubation behavior and care of young: is there a causal relationship? Ann. NY Acad. Sci. 474:252–267.

Butler AB, Reiner A, Karten HJ. 2012. Evolution of the amniote pallium and the origins of mammalian neocortex. Ann. NY Acad. Sci. 1225:14–27.

Butte NF. 2000. Carbohydrate and lipid metabolism in pregnancy: normal compared with gestational diabetes mellitus. Am. J. Clin. Nutr. 71:1256S–1261S.

Cabanac M. 1987. The place of behaviour in physiology. in biology of stress in farm animals: an integrative approach. In Wiepkema PR, Van Adrichem PWM. (Eds.) *Current topics in veterinary medicine and animal science* (pp. 187–194). Vol 42. Dordrecht: Springer.

Cabanac M, Duclaux R, Spector NH. 1971. Sensory feedback in regulation of body weight: is there a ponderostat? Nature 229:125–127.

Cabanac M, Massonnet B. 1974. Temperature regulation during fever: change of set point or change of gain? A tentative answer from a behavioural study in man. J. Physiol. 238:561–568.

Cannon WB. 1963. *Wisdom of the body*. New York: WW Norton and Company Inc.

Cao C, Matsumura K, Shirakawa N, Maeda M, Jikihara I, Kobayashi S, Watanabe Y. 2001. Pyrogenic cytokines injected into the rat cerebral ventricle induce cyclooxygenase-2 in brain endothelial cells and also upregulate their receptors. Eur. J. Neurosci. 13:1871–1790.

Carswell EA. Old LJ, Kassel RL, Green S, Fiore N, Williamson B. 1975. An endotoxin-induced serum factor that causes necrosis of tumors. Proc. Natl. Acad. Sci. 72:3666–3670.

Catalano P. 2002. The diabetogenic state of maternal metabolism in pregnancy. NeoReviews 3:e165.

Chamberlin TC. 1890. The method of multiple working hypotheses. Science 15:92–96.

Chao HM, Choo PH, McEwen BS. 1989. Glucocorticoid and mineralocorticoid receptor mRNA expression in rat brain. Neuroendocrinol. 50:365–371.

Cheng MF. 1975. Induction of incubation behaviour in male ring doves (*Streptopelia risorii*): a behavioural analysis. J. Reprod. Fertil. 42:267–276.

Cheng MF, Silver R. 1975. Estrogen-progesterone regulation of nest-building and incubation behavior in ovariectomized ring doves (*Streptopelia risoria*). J. Comp. Physiol. Psychol. 88:256–263.

Chmura HE, Duncan C, Saer B, Moore JT, Barnes BM, Buck CL, Christian HC, Loudon AS, Williams CT. 2022. Hypothalamic remodelling of thyroid hormone signalling during hibernation in the arctic ground squirrel. Commun. Biol. 5:492.

Clark JT, Kalra PS, Crowley WR, Kalra SP. 1984. Neuropeptide Y and human pancreatic polypeptide stimulate feeding behaviour in rats. Endocrinol. 115:427–429.

Clarkson J, Han SY, Piet R, McLennan T, Kane GM, Ng J, Porteous RW, Kim JS, College WH, Iremonger KJ, Herbison AE. 2017. Definition of the hypothalamic GnRH pulse generator in mice. Proc. Natl. Acad. Sci. 114:E10216–E10223.

Cockburn A. 2006. Prevalence of different modes of parental care in birds. Proc. R. Soc. B 273:1375–1383.

Cohen P, Zhao C, Cai X, Montez JM, Rohani CS, Feinstein P, Mombaerts P, Friedman JM. 2001. Selective deletion of leptin receptor in neurons leads to obesity. J. Clin. Invest. 108:1113–1121.

Colin J, Timbal J, Boutelier C, Houdas Y, Siffre M. 1968. Rhythm of the rectal temperature during a 6-month free-running experiment. J. Appl. Physiol. 25:170–176.

Cone RD, 2005. Anatomy and regulation of the central melanocortin system. Nat. Neurosci. 8:571–578.

Contreras-Alcantara S, Baba K, Tosini G. 2010. Removal of melatonin receptor type 1 induces insulin resistance in the mouse. Obesity 18:1861–1863.

Coxe JR. 1846. *The writings of Hippocrates and Galen*. Philadelphia: Lindsay and Blakiston.

Christou F, Pitteloud N, Gomez F. 2017. The induction of ovulation by pulsatile administration of GnRH: an appropriate method in hypothalamic amenorrhea. Gynecol. Endocrinol. 33:598–601.

Corner GW, Allen WM. 1929. Physiology of the corpus luteum, II: production of a special uterine reaction (progestational proliferation) by extracts of the corpus luteum. Amer. J. Physiol. 88:326–339.

Cunningham FJ, Wilson SC, Knight PG, Gladwell RT. 1984. Chicken ovulation cycle. J. Exp. Zool. 232:485–494.

Cuvier G. 1805. *Lecons d'anatomie comparee*. 5 vols. Paris: Baudouin.

Czieselsky K, Prescott M, Porteous R, Campos P, Clarkson J, Steyn FJ, Campbell RE, Herbison AE. 2016. Pulse and surge profiles of luteinizing hormone secretion in the mouse. Endocrinol. 157:4794–4802.

Czeisler CA, Duffy JF, Shanahan TL, Brown EN, Mitchell JF, Rimmer DW, Ronda JM, Silva EJ, Allan JS, Emens JS, Dijk DJ, Kronauer RE. 1999. Stability, precision, and near-24-hour period of the human circadian pacemaker. Science 284:2177–2181.

D'Alecy LG, Kluger MJ. 1975. Avian febrile response. J. Physiol. 253:223–232.

Daan S. 2010. A history of chronobiological concepts. In Albrecht U. (Ed.) *The circadian clock* (pp. 1–35). New York: Springer Publishers.

Dallman MF, Jones MT. 1973. Corticosteroid feedback control of ACTH: effect of stress-induced corticosterone secretion on subsequent stress responses in the rat. Endocrinol. 92:1367–1375.

Dardente H, Hazlerigg DG, Ebling FJP. 2014. Thyroid hormone and seasonal rhythmicity. Front. Endocrinol. 5:19.

Dark J, Pickard GE, Zucker I. 1985. Persistence of circannual rhythms in Ground squirrels with lesions of the suprachiasmatic nuclei. Brain Res. 332:201–207.

Darwin C, Darwin F. 1898. *The power of movement in plants*. New York: D Appleton and Co.

de Creccio L. 1865. Sopra un caso di apparenze virili in una donna. Il Morgagni. 7:151–189.

De Gorter J. 1736. *De perspiratione insensibili*. Batavorum: Vander.

de Mairan, J. 1729. Observation botanique. Hist. Acad. Roy. Sci. 35–36.

Dedovic K, Renwick R, Mahani NK, Engert V, Lupien SJ, Pruessner JC. 2005. The Montreal Imaging Stress Task: using functional imaging to investigate the effects of perceiving and processing psychosocial stress in the human brain. J. Psychiatry Neurosci. 30:319–325.

Dedovic K, Rexroth M, Wolff E, Duchesne A, Scherling C, Beaudry T, Lue SD, Lord C, Engert V, Pruessner JC. 2009. Neural correlates of processing stressful information: an event-related fMRI study. Brain Res. 1293:49–60.

Del Giudice EM, Santoro N, Cirillo G, D'Urso L, Di Toro R, Perrone L. 2001. Mutational screening of the CART gene in obese children: identifying a mutation (Leu34Phe) associated with reduced resting energy expenditure and cosegregating with obesity phenotype in a large family. Diabetes 50:2157–2160.

Dempsey PW, Vaidya SA, Cheng G. 2003. The art of war: innate and adaptive immune responses. Cell Mol. Life Sci. 60:2604–2621.

Dinarello CA, Renfer L, Wolff SM. 1977. Human leukocytic pyrogen: purification and development of a radioimmunoassay. Proc. Natl. Acad. Sci. 74:4624–4627.

Dopico XC, Evangelou M, Ferreira RC, Guo H, Pekalski ML, Smyth DJ, Cooper N, Burren OS, Fulford AJ, Hennig BJ, Prentice AM, Ziegler AG, Bonifacio E, Wallace C, Todd JA. 2015. Widespread seasonal gene expression reveals annual differences in human immunity and physiology. Nature Commun. 6:7000.

Doi O, Takai T, Nakamura T, Tanabe Y. 1980. Changes in the pituitary and plasma LH plasma and follicular progesterone and estradiol, and plasma testosterone and estrone concentrations during the ovulatory cycle of the quail (Coturnix japonica). Gen Comp. Endocrinol. 41:156–163.

Drickamer LC, 1990.Seasonal variation in fertility, fecundity and litter sex ratio in laboratory and wild stocks of house mice (*Mus domesticus*). Lab. Anim. Sci. 40:284–288.

Dunn IC, Beattie KK, Maney D, Sang HM, Talbot RT, Wilson PW, Sharp PJ. 1996. Regulation of chicken gonadotropin-releasing hormone-I mRNA in incubating, nest-deprived and laying Bantam hens. Neuroendocrinol. 63:504–513.

Ebihara S, Ushiyama K, Oshima I. 1984. Circadian organization in the pigeon, *Columbia livia*: the role of the pineal organ and the eye. J. Comp. Physiol. 154:59–69.

Ebling FJP. 2015. Hypothalamic control of seasonal changes in food intake and body weight. Front. Neuroendocrinol. 37:97–107.

Ebling FJP, Barrett P. 2008. The regulation of seasonal changes in food intake and body weight. J. Neuroendocrinol. 20:827–833.

Edgar RS, Green EW, Zhao Y, van Ooijen G, Olmedo M, Qin X, Xu Y, Pan M, Valekunja UK, Feeney KA, Maywood ES, Hastings MH, Baliga NS, Merrow M, Millar AJ, Johnson CH, Kyriacou CP, O'Neill JS, Reddy AB. 2012. Nature 485:459–464.Edgar RS, Stangherlin A, Nagy AD, Reddy AB. 2016. Cell autonomous regulation of herpes and influenza virus infection by the circadian clock. Proc. Natl. Acad. Sci. 133:10085–10090.El Halawani ME, Burke WH, Dennison PT, 1980. Effect of nest-deprivation on serum prolactin levels in nesting female turkeys. Biod. Reprod. 23:118–123.

Elias CF, Lee CE, Kelly JF, Ahima RS, Kuhar M, Saper CB, Elmquist JK. 2001. Characterization of CART neurons in the rat and human hypothalamus. J. Comp. Neurol. 432:1–19.

Elmquist JK, Bjorbaek C, Ahima RS, Flier JS, Saper CB. 1998. Distributions of leptin receptor mRNA isoforms in the rat brain. J. Comp. Neurol. 395:535–547.

Elmquist JK, Elias CF, Saper CB. 1999. From lesions to leptin: hypothalamic control of food intake and body weight. Neuron 22:221–232.Elphinstone RE, Weckman AM, McDonald CR, Tran V, Zhong K, Madanista M, Kalilani-Phiri L, Khairallah C, Taylor SM, Meshnick SR, Mwapasa V, O Ter Kuile F, Conroy AL, Kain KC. 2019. Early malaria infection, dysregulation of angiogenesis, metabolism and inflammation across pregnancy, and risk of preterm birth in Malawi: a cohort study. PLoS Med. 16:e1002914.

Etches RJ, Petitte JN. 1990. Reptilian and avian follicular hierarchies: models for the study of ovarian development. J. Exp. Zool. 4:112–122.

Evans SS, Repasky EA, Fisher DT. 2015. Fever and the thermal regulation of immunity: the immune system feels the heat. Nat. Rev. Immunol. 15:335–349.

Everest P, Wain J, Roberts M, Rook G, Dougan G. 2001. The molecular mechanisms of severe typhoid fever. Trends Microbiol. 9:316–320.

Farr A, Chervenak FA, McCullough LB, Baergen RN, Grunebaum A. 2018. Human placentophagy: a review. Am. J. Obstet Gynecol. 218:401.

Fellner OO. 1912. Experimentall erzeugte Wachstumsveranderungen am weiblichen Genitale der Kaninchen. Centralbl. Allg. Path. Anat. 23:673–676.

Fleming AS, O'Day DH, Kraemer GW. 1999. Neurobiology of mother-infant interactions: experience and central nervous system plasticity across development and generations. Neurosci. Biobehav. Rev. 23:673–685.

Follett BK. 2015. Seasonal changes in the neuroendocrine system: some reflections. Front. Neuroendocrinol. 37:3–12.

Fonken LK, Workman JL, Walton JC, Nelson RJ. 2010. Light at night increases body mass by shifting the time of food intake. Proc. Natl. Acad. Sci. 107:18664–18669.

Forest JC, Garrido-Russo M, Lemay A, Carrier R, Dube JL. 1983. Reference values for the oral glucose tolerance test at each trimester of pregnancy. Amer. J. Clin. Pathol. 80:828–831.

Foster RG, Roenneberg T. 2008. Human responses to the geophysical daily, annual and lunar cycles. Curr. Biol. 18:R784–R794.

Foster RG, Follett BK, Lythgoe JN. 1985. Rhodopsin-like sensitivity of extra-retinal photoreceptors mediating the photoperiodic response in quail. Nature 313:50–52.

Francis D, Diorio J, Liu D, Meaney MJ. 1999. Nongenomic transmission across generations of maternal behavior and stress responses in the rat. Science 286:1155–1158.

Freedman MS, Lucas RJ, Soni B, von Schantz M, Munoz M, David-Gray Z, Foster RG. 1999. Regulation of mammalian circadian behavior by non-rod, non-cone ocular photoreceptors. Science 284:502–504.

Friedman MH, Lapham M. 1931. A simple, rapid procedure for the diagnosis of early pregnancies. Am. J. Obs. Gynae. 21:405–410.

Furay AR, Bruestle AE, Herman JP. 2008. The role of the forebrain glucocorticoid receptor in acute and chronic stress. Endocrinol. 149:5482–5490.

Gamble KL, Berry R, Frank SJ, Young ME. 2014. Circadian clock control of endocrine factors. Nat. Rev. Endocrinol. 10:466–475.

Gerra G, Zaimovic A, Mascetti GG, Gardini S, Zambelli U, Timpano M, Raggi MA, Brambilla F. 2001. Neuroendocrine responses to experimentally induced psychological stress in healthy humans. Psychoneuroendocrinol. 26:91–107.

Ghezzi I, Ruggles C. 2007. Chankillo: a 2300-year-old solar observatory in coastal Peru. Science 315:1239–1243.

Gibson MJ, Wu TJ, Miller GM, Silverman AJ. 1997. What nature's knockout teaches us about GnRH activity: hypogonadal mice and neuronal grafts. Horm. Behav. 31:212–220.

Gobler CJ, Doherty OM, Hattenrath-Lehmann TK, Griffith AW, Kang Y, Litaker RW. 2017. Ocean warming since 1982 has expanded the niche of toxic algal blooms in the North Atlantic and North Pacific Oceans. Proc. Natl. Acad. Sci. 114:4975–4980.

Goldsmith AR, Burke S, Prosser JM. 1984. Inverse changes in plasma prolactin and LH concentrations in female canaries after disruption and reinitiation of incubation. J. Endocrinol. 103:251–256.

Goodman RL, Legan SJ, Ryan KD, Foster DL, Karsch FL. 1981. Importance of variation in behavioural and feedback actions of oestradiol to the control of seasonal breeding in the ewe. J. Endocrinol. 89:229–240.

Gore AC. 2002. *GnRH: the master molecule of reproduction*. New York: Kluwer Academic Publishers.

Gray GD, Soderstein P, Tallentire D, Davidson JM. 1978. Effects of lesions in various structures of the suprachiasmatic-preoptic region on LH regulation and sexual behavior in female rats. Neuroendocrinol. 25:174–191.

Greco CM, Koronowski KB, Smith JB, Shi J, Kunderfranco P, Carriero R, Chen S, Samad M, Welz PS, Zinna VM, Mortimer T, Chun SK, Shimaji K, Sato T, Petrus P, Kumar A, Vaca-Dempere M, Deryagin O, Van C, Kuhn JMM, Lutter D, Seldin M, Masri S, Li W, Baldi P, Dyar K, Munoz-Canoves P, Benitah SA, Sassone-Corsi P. 2021. Integration of feeding behavior by the liver circadian clock reveals network dependency of metabolic rhythms. Sci. Advances 7:eabi7828.

Gropp E, Shanabrough M, Borok E, Xu AW, Janoschek R, Buch T, Plum L, Balthasar N, Hampel B, Waisman A, Barsh GS, Horvath TL, Bruning JC. 2005. Agouti-related peptide-expressing neurons are mandatory for feeding. Nat. Neurosci. 8:1289–1291.

Gwinner E, Benzinger I. 1978. Synchronization of a circadian rhythm in pinealectomized European starlings by daily injections of melatonin. J. Comp. Physiol. 127:209–213.

Gwinner E, Dittami J. 1990. Endogenous reproductive rhythms in a tropical bird. Science 249:906–908.

Hardin PE, Hall JC, Rosbash M. 1990. Feedback of the *Drosophila* period gene product on circadian cycling of its messenger RNA levels. Nature 343:536–540.

Halberg F, Engeli M, Hamburger C, Hillman D. 1965. Spectral resolution of low-frequency, small-amplitude rhythms in excreted 17-ketosteroid: probable androgen induced circaseptan desynchronization. Acta. Endocrinol. 103:5–54.

Halberg F, Lagoguey M, Reinberg A. 1983. Human circannual rhythms over a broad spectrum of physiological processes. Inter. J. Chronobiol. 8:225–268.

Handwerger S, Freemark M. 2000. The roles of placental growth hormone and placental lactogen in the regulation of human fetal growth and development. J. Pediatr. Endocrinol. Metab. 13:343–356.

Harris GW. 1948. Neural control of the pituitary gland. Physiol. Reb. 28:139–179.

Hastings MH, Maywood ES, Brancaccio M. 2018. Generation of circadian rhythms in the suprachiasmatic nucleus. Nat. Rev. Neurosci. 19:453–469.

Hau M, Wikelsko M, Wingfield JC. 1998. A neotropical forest bird can measure the slight changes in tropical photoperiod. Proc. R. Soc. Lond. B 265:89–95.

Hebb DO. 1949. *The organization of behavior*. New York: Wiley & Sons.

Helfer G, Ross AW, Thomson LM, Mayer CD, Stoney PN, McCaffery PJ, Morgan PJ. 2016. A neuroendocrine role for chemerin in hypothalamic remodelling and photoperiodic control of energy balance. Sci. Rep. (6):26830.

Helfer G, Stevenson TJ. 2020. Pleiotropic effects of proopiomelanocortin and VGF nerve growth factor inducible neuropeptides for the long-term regulation of energy balance. Molecular and Cellular Endocrinology 514:110876.

Helm B, Van Doren BM, Hoffmann D, Hoffmann U. 2019. Evolutionary response to climate change in migratory Pied flycatchers. Curr. Biol. 29:3714–3719.

Herbison AE. 1998. Multimodal influence of estrogen upon gonadotropin-releasing hormone neurons. Endoc. Rev. 19:302–330.

Herbison AE. 2016. Control of puberty onset and fertility by gonadotropin-releasing hormone neurons. Nat. Rev. Endocrinol. 12:452–466.

Herman JP, Watson SJ, Chao HM, Coirini H, McEwen BS. 1993. Diurnal regulation of glucocorticoid receptor and mineralocorticoid receptor mRNA in rat hippocampus. Mol. Cell. Neurosci. 4:181–190.

Hinde RA. 1962. Temporal relations of brood patch development in domesticated canaries. IBIS 104:90–97.

Hinkle PM, Tashjian Jr AH. 1975. Thyrotropin-releasing hormone regulates the number of its own receptors in the GH3 strain of pituitary cells in culture. Biochem. 14:3845–3851.

Hirose T. 1920. Exogenous stimulation of corpus luteum formation in the rabbit: influence of extracts of human placenta, decidua, fetus, hydatid mole, and corpus luteum on the rabbit gonad. J. Jpn. Gynecol. Soc. 16:1055.

Hodgkin AL, Huxley AF. 1952. A quantitative description of membrane current and its application to conduction and excitation in nerve. J. Physiol. 117:500–544.

Hopfield JJ. 1982. Neural networks and physical systems with emergent collective computational abilities. Proc. Natl. Acad. Sci. 79:2554–2558.

Hunzicker-Dunn M, Mayo K. 2015. Gonadotropin signaling in the ovary. In Tony MP, Anthony JZ (Eds.) *Knobil and Neill's physiology of reproduction* (pp. 895–945). 4th Ed. Academic Press.

Huppertz B. 2008. The anatomy of the normal placenta. J. Clin. Pathol. 61:1296–1302.

Ingalls AM, Dickie MM, Snell GD. 1950. Obesity, a new mutation in the house mouse. J. Hered. 41:317–318.

IPCC, 2022: Summary for Policymakers. In Pörtner H-O, Roberts DC, Tignor M, Poloczanska ES, Mintenbeck K, Alegría A, Craig M, Langsdorf S, Löschke S, Möller V, Okem A, Rama B. (Eds.) *Climate Change 2022: Impacts, adaptation and vulnerability. contribution of working group ii to the sixth assessment report of the Intergovernmental Panel on Climate Change* (pp. 3–33). Cambridge, UK/New York: Cambridge University Press.

Janeway Jr CA, Medzhitov R. 2002. Innate immune regulation. Ann. Rev. Immunol. 20:197–216.

Jethwa PH, Warner A, Fowler MJ, Murphy M, de Backer MW, Adan RA, Barrett P, Brameld JM, Ebling FJ. 2010. Short-days induce weight loss in Siberian hamsters despite overexpression of the agouti-related peptide gene. J. Neuroendocrinol. 22:564–575.

Johns JE, Pfeiffer EW. 1963. Testosterone-induced incubation patches of phalarope birds. Science 140:1225–1226.

Johns AL, Johnson PA, van Tienhoven A. 1984. Ovulatory response, and plasma concentrations of luteinizing hormone and progesterone following administration of synthetic mammalian or chicken luteinizing hormone releasing hormone relative to the first of second ovulation in the sequence of the domestic hen. Biol. Reprod. 31:646–655.

Johnson AL. 1986. Reproduction in the female. In Sturkie PD. (Ed.) *Avian physiology* (pp. 403–431). New York: Springer-Verlag.

Johnson AL, Johnson PA, van Tienhoven A. 1984. Ovulatory response, and plasma concentrations of luteinizing hormone and progesterone following administration of synthetic mammalian or chicken luteinizing hormone-releasing hormone relative to the first or second ovulation in the sequence of the domestic hen. Biol. Reprod. 31:646–655.

Kalliolias GD, Ivashkiv LB. 2016. TNF biology, pathogenic mechanisms and emerging therapeutic strategies. Nat. Rev. Rheumatol. 12:49–62.

Kalsbeek A, Strubbe JH. 1998. Circadian control of insulin secretion is independent of the temporal distribution of feeding. Physiol. Behav. 63:553–560.

Kameda Y, Arai Y, Nishimaki T. 2003. Ultrastructural localization of vimentin immunoreactivity and gene expression in tanycytes and their alterations in hamsters kept under different photoperiods. Cell Tissue Res. 314:251–62.

Karl A, Schaefer M, Malta LS, Dorfel D, Rohleder N, Werner A. 2006. A meta-analysis of structural brain abnormalities in PTSD. Neurosci. Biobehav Rev. 30:1004–1031.

Katz PS. 2016. Evolution of central pattern generators and rhythmic behaviours. Phil. Trans. R. Soc. B. 371:20150057.

Kenagy GJ. 1980. Interrelation of endogenous annual rhythms of reproduction and hibernation in the golden-mantled ground squirrel. J. Comp. Physiol. 135:333–339.

Kimbrell DA, Beutler B. 2001. The evolution and genetics of innate immunity. Nat. Rev. Genet. 2:256–267.

Kimura M, Toth LA, Agostini H, Cady AB, Majde JA, Krueger JM. 1994. Comparison of acute phase responses induced in rabbits by lipopolysaccharide and double stranded RNA. Am. J. Physiol. 267:R1596–R1605.

Kirschbaum C, Pirke K.M. Hellhammer D.H. 1993. The "Trier Social Stress Test"—a tool for investigating psychobiological stress responses in a laboratory setting. Neuropsychol. 28:76–81.

Kitamura T, Kahn CR, Acili D. 2003. Insulin receptor knockout mice. Ann. Rev. Physiol. 65:313–332.

Kluger MJ. 1978. The evolution and adaptive value of fever. Am. Sci. 66:38–43.

Kluger MJ. 1979. *Fever: its biology, evolution and function*. Princeton, NJ: Princeton University Press.

Konopka RJ, Benzer S. 1971. Clock mutants of *Drosophila melanogaster*. Proc. Natl. Acad. Sci. 68:2112–2116.

Kozak W, Conn CA, Kluger MJ. 1994. Lipopolysaccharide induces fever and depresses locomotor activity in unrestrained mice. Am. J. Physiol 266: R125–R135.

Krashes MJ, Koda S, Ye C, Rogan SC, Adams AC, Cusher DS, Maratos-Flier E, Roth BL, Lowell BB. 2011. Rapid, reversible activation of AgRP neurons drives feeding behaviour in mice. J. Clin. Invest. 121:1424–1428.

Kristensen P, Judge ME, Thim L. 1998. Hypothalamic CART is a new anorectic peptide regulated by leptin. Nature 393:72–76.

Kristensen P, Judge ME, Thim L, Ribel U, Christjansen KN, Wulff BS, Clausen JT, Jensen PB, Madsen OD, Vrang N, Larsen PJ, Hastrup S. 1998. Hypothalamic CART is a new anorectic peptide regulated by leptin. Nature. 393:72–76.

Kojima M, Hosoda H, Date Y, Nakazato M, Matsuo H, Kangawa K. 1999. Ghrelin is a growth-hormone-releasing acylated peptide from stomach. Nature 402:656–660.

Kreitzman L, Foster RG 2004. *The rhythms of life: the biological clocks that control the daily lives of every living thing*. New Haven: Yale University Press.

Kuenzel WJ. 1999. A neural system proposed to regulate the annual cycle of migratory birds. In: Adams NJ, Slotow RH (Eds.) *Proceedings of the 22nd International Ornithology Congress* (pp. 784–792). Durban/Johannesburg: Birdlife South Africa

Kuenzel WJ, Douglass LW, Davidson BA. 1987. Robust feeding following central administration of neuropeptide Y or peptide YY in chicks, *Gallus domesticus*. Peptides 8:823–828.Lack D. 1968. *Ecological adaptations for breeding in birds*. London: Methuen.

Ladyman SR, Brooks VL. 2021. Central actions of insulin during pregnancy and lactation. J. Neuroendocrinol. 33:e12946.

Lamia KA, Papp SJ, Yu RT, Barish GD, Uhlenhaut NH, Jonker JW, Downes M, Evans RM. 2011. Cryptochromes mediate rhythmic repression of the glucocorticoid receptor. Nature 480:552–556.

Lamia KA, Storch KF, Weitz CJ. 2008. Physiological significance of a peripheral tissue circadian clock. Proc. Natl. Acad. Sci. 105:15172–15177.Lapatto R, Pallais JC, Zhang D, Chan YM, Mahan Am, Cerrato F, Le WW, Hoffman GE, Seminara SB. 2007. Kiss1-/- mice exhibit more variable hypogonadism than Gpr54-/- mice. Endocrinol. 148:4927–4936.

Lawlor DA, Relton C, Sattar N, Nelson SM. 2012. Maternal adiposity—a determinant of perinatal and offspring outcomes? Nat. Rev. Endocrinol. 8:679–688.

Lazarus M, Yoshida K, Coppari R, Bass CE, Mochizuki T, Lowell BB, Saper CB. 2007. EP3 prostaglandin receptors in the median preoptic nucleus are critical for fever responses. Nat. Neurosci. 10:1131–1133.

Lazutkaite G, Solda A, Lossow K, Meyerhof W, Dale N. 2017. Amino acid sensing in hypothalamic tanycytes via umami taste receptors. Mol. Metab. 6:1480–1492.

Lee DA, Bedont JL, Pak T, Wang H, Song J, Miranda-Angulo A, Takiar V, Charubhumi V, Balordi F, Takebayashi H, Aja S, Ford E, Fishell G, Blackshaw S. 2012. Tanycytes of the hypothalamic median eminence form a diet-responsive neurogenic niche. Nat. Neurosci. 15:700–702.

Lee JH, Miele ME, Hicks DJ, Phillips KK, Trent JM, Weissman BE, Welch DR. 1996. KiSS-1, a novel human malignant melanoma metastasis-suppressor gene. J. Natl. Cancer Inst. 88:1731–1737.

Lee YS, Challis BG, Thompson DA, Yeo GS, Keogh JM, Madonna ME, Wraight V, Simms M, Vatin V, Meyre D, Shield J, Burren C, Ibrahim Z, Cheetham T, Swift P, Blackwood A, Hung CC, Wareham NJ, Froguel P, Millhauser GL, O'Rahilly S, Farooqi IS. 2006. A POMC variant implicates β-melanocyte-stimulating hormone in the control of human energy balance. Cell Metab. 3:135–140.

Lehrman DS. 1964. The reproductive behavior of ring doves. Sci Amer. 211:48–55.Lehrman DS, Brody PN. 1964. Effect of prolactin on established incubation behavior in the ring dove. J. Comp. Physiol. Psych. 57:161–165.Leliavski A, Shostak A, Husse J, Oster H. 2014. Impaired glucocorticoid production and response to stress in Arntl-deficient male mice. Endocrinol. 155:133–142.

Leng G, Brown D. 1997. The origins and significance of pulsatility in hormone secretion from the pituitary. J. Neuroendocrinol. 9:493–513.

Leranth C, MacLusky NJ, Sakamoto H, Shanabrough M, Naftolin F. 1985. Glutamic acid decarboxylase containing axons synapse on LHRH neurons in the rat medial preoptic area. Neuroendocrinol. 40:536–539.

Lewis JE, Ebling FJP. 2018. Hamsters as model species for neuroendocrine studies. In Ludwig M, Levkowitz G. (Eds.) *Model animals in neuroendocrinology: from worm to mouse to man* (pp. 161–180). Oxford, UK: Wiley & Sons Ltd.

Lightman SL, Conway-Campbell BL. 2010. The crucial role of pulsatile activity of the HPA axis for continuous dynamic equilibration. Nat. Rev. Endocrinol. 11:710–718.

Lincoln GA. 2019. A brief history of circannual time. J. Neuroendocrinol. 31:e12694.

Lincoln GA, Clarke IJ, Hut RA, Hazlerigg DG. 2006. Characterizing a mammalian circannual pacemaker. Science 314:1941–1944.

Liu HK, Long DW, Bacon LW. 2001. Preovulatory luteinizing hormone surge interval in old and young laying turkey hens early in the egg production period. Poultry Sci. 80:1364–1370.

Lombardo MF, De Angelis F, Bova L, Bartolini B, Bertuzzi F, Nano R, Capuani B, Lauro R, Federici M, Lauro D, Donadel G. 2011. Human placental lactogen (hPLA) activates signalling pathways linked to cell survival and improves insulin secretion in human pancreatic islets. Islets 3:250–258.

Lovallo WR, Farag NH, Vincent AS. 2010. Use of a resting control day in measuring the cortisol response to mental stress: diurnal patterns, time of day and gender effects. Psychoneuroendocrinol. 35:1253–1258.

MacDougall-Shackleton SA, Stevenson TJ, Watts HE, Pereyra ME, Hahn TP, 2009. The evolution of photoperiod response systems and seasonal GnRH1 plasticity in birds. Inte. Comp. Biol. 49:580–589.

MacMicking JD. 2012. Interferon-inducible effector mechanisms in cell-autonomous immunity. Nat. Rev. Immunol. 12:367–382.

Maletinska L, Maixnerova J, Matyskova R, Haugvicova R, Pirnik Z, Kiss A, Zelezna B. 2008. Synergistic effect of CART (cocaine- and amphetamine-regulated transcript) peptide and cholecystokinin on food intake regulation in lean mice. BMC Neurosci. 9:101.

Malville JM, Wendorf F, Mazar AA, Schild R. 1998. Megaliths and neolithic astronomy in southern Egypt. Nature 392:488–491.

Marcinkevage JA, Venkat Narayan KM. 2010. Gestational diabetes mellitus: taking it to heart. Prim. Care Diabetes 5:81–88.

Marder E, Bucher D. 2001. Central pattern generators and the control of rhythmic movements. Curr. Biol. 11:R986-R996.

Marks JL, Porte Jr. D, Stahl WL, Baskin DG. 1990. Localization of insulin receptor mRNA in rat brain by in situ hybridization. Endocrinol. 127:3234–3236.

Martin P, Cortina-Borja M, Newburn M, Harper G, Gibson R, Dodwell M, Dattani N, Macfarlane A. 2018. Timing of singleton births by onset of labour and mode of birth in NHS maternity units in England, 2005–2014: a study of linked birth registration, birth notification, and hospital episode data. PLoS One 13(6): e0198183.

Martinez de la Escalera G, Choi ALH, Weiner RI. 1992. Generation and synchronization of gonadotropin-releasing hormone pulses: intrinsic properties of the GT1-1 GnRH neuronal cell line. Proc. Natl. Acad. Sci. 89:1852–1855.

Martins BMC, Tooke AK, Thomas P, Locke JCW. 2018. Cell size control driven by the circadian clock and environment in cyanobacteria. Proc. Natl. Acad. Sci. 115:E11415–E11424.

Masferrer JL, Zweifel BS, Manning PT, Hauser SD, Leahy KM, Smith WG, Isakson PC, Seibert K. 1994. Selective inhibition of inducible cyclooxygenase 2 in vivo is anti-inflammatory and nonulcerogenic. Proc. Natl. Acad. Sci. 91:3228–3232.

Maxwell JC. 1868. On governors. Proc. R. Soc. London 16:270–283.

Mayr O. 1970. *The origins of feedback control.* Clinton, MA: The Colonial Press.

McCarthy MM. 2008. Estradiol and the developing brain. Physiol. Rev. 88:91–124.

McEwen BS. 1998. Protective and damaging effects of stress mediators. New Engl. J. Med. 338:171–179.

McEwen BS. 2000. Allostasis and allostatic load: implications for neuropsychopharmacology. Neuropsychoparmacol. 22:108–124.

McEwen BS, Lasley EN. 2002. *The end of stress as we know it.* Washington, DC: Joseph Henry Press.

McEwen BS, Wingfield JC. 2003. The concept of allostasis in biology and biomedicine. Horm. Behav. 43:2–15.

McShane TM, Petersen SL, McCrone S, Keisler DH. 1993. Influence of food restriction on neuropeptide-Y, proopiomelanocortin, and luteinzing hormone releasing hormone gene expression in sheep hypothalamic. Biol. Reprod. 49:831–839.

Meaney MJ. 2001. Maternal care, gene expression, and the transmission of individual differences in stress reactivity across generations. Ann. Rev. Neurosci. 24:1161–1192.

Mellon PL, Windle JJ, Goldsmith PC, Padula CA, Roberts JL, Weiner RI. 1990. Immortalization of hypothalamic GnRH neurons by genetically targeted tumorigenesis. Neuron 5:1–10.

Messager S, Chatzidaki EE, Ma D, Hendrick AG, Zahn D, Dixon J, Thresher RR, Malinge I, Lomet D, Carlton MBL, College WH, Caraty A, Aparicio SAJR. 2005. Kisspeptin directly stimulates gonadotropin-releasing hormone release via G protein-coupled receptor 54. Proc. Natl. Acad. Sci. 102:1761–1766.

Meyer C, Muto V, Jaspar M, Kusse C, Lambot E, Chellappa SL, Degueldre C, Balteua E, Luxen A, Middleton B, Archer SN, Collette F, Dijk DJ, Phillips C, Maquet P, Vandewalle G. 2016. Seasonality in human cognitive brain responses. Proc. Natl. Acad. Sci. 113:3066–3071.

Migaud M, Butrille L, Batailler M. 2015. Seasonal regulation of structural plasticity and neurogenesis in the adult mammalian brain: focus on the sheep hypothalamus. Front. Neuroendocrinol 37:146–57.

Miller BH, Olson SL, Turek FW, Levine JE, Horton TH, Takahashi JS. 2004. Circadian clock mutation disrupts estrous cyclicity and maintenance of pregnancy. Curr. Biol. 14:1367–1373.

Miller WL. 2013. A brief history of adrenal research: steroidogenesis—the soul of the adrenal. Mol. Cell. Endocrinol. 371:5–14.

Miner JL, Della-Fera MA, Paterson JA, Baile BA. 1989. Lateral cerebroventricular injections of neuropeptide Y stimulates feeding in sheep. Am. J. Physiol. 257: R383–R387.

Minorsky N. 1922. Directional stability of automatically steered bodies. J. Amer. Soc. Naval Engineers 34:280–309.

Moore AM, Coolen LM, Porter DT, Goodman RL, Lehman MN. 2018. KNDy cells revisited. Endocrinol. 159:3219–3234.

Moore RY, Eichler VB. 1972. Loss of circadian adrenal corticosterone rhythm following suprachiasmatic lesions in the rat. Brain Res. 42:201–206.Morley JE, Hernandez EN, Flood JF. 1987. Neuropeptide Y increases food intake in mice. Am. J. Physiol. 253: R516–R522.

Morris CJ, Purvis TE, Mistretta J, Scheer FAJL. 2016. Effects of the internal circadian system and circadian misalignment on glucose tolerance in chronic shift workers. J. Clin. Endo. Metab. 101:1066–1074.

Morris CJ, Yang JN, Garcia JI, Myers S, Bozzi I, Wang W, Buxton OM, Shea SA, Scheer FAJL. 2015. Endogenous circadian system and circadian misalignment impact glucose tolerance via separate mechanisms in humans. Proc. Natl Acad. Sci. 112:E2225–E2234.Mrosovsky N. 1990. *Rheostasis: the physiology of change*. New York: Oxford University Press.

Murphy K, Weaver C. 2016. *Janeway's immunobiology*. New York: Garland Science Taylor and Francis Group.

Murton RK, Westwood NJ. 1977. *Avian breeding cycles*. Oxford: Oxford University Press.

Nader N, Chrousos GP, Kino T. 2010. Interactions of the circadian CLOCK system and the HPA axis. Trends Endocinol. Metabol. 21:277–286.

Nahum GG, Huffaker BJ. 1990. Correlation between first- and early third-trimester glucose screening test results. Obstet. Gynecol. 76:709–713.

Nakazato M, Murakami N, Date Y, Kojima M, Matsuo H, Kangawa K, Matsukura S. 2001. A role for ghrelin in the central regulation of feeding. Nature 409:194–198.

Needham J. 1959. *A history of embryology*. Cambridge, UK: Cambridge University Press.

Nicolaidis S. 1977. Physiologie du comportement alimentaire. In Meyer P (Ed.) *Physiologie humaine* (pp. 908–922). Paris: Flammarion Medecine-Sciences.

Nomoto S. 1996. Diurnal variation in fever induced by intravenous LPS injection in pigeons. Pflugers Arch. Eur. J. Physiol. 431:987–989.

Ollmann NM,Wilson BD, Yang YK, Kerns JA, Chen Y, Gantz I, Barsh GS. 1997. Antagonism of central melanocortin receptors in vitro and in vivo by agouti-related protein. Science 278:135–138.

O'Neill JS, Reddy AB. 2011. Circadian clocks in human red blood cells. Nature 469:498–503.

O'Neill LAJ, Golenbock D, Bowie AG. 2013. The history of Toll-like receptors—redefining innate immunity. Nat Rev Immunol. 13:453–460.

Orellana JA, Saez PJ, Cortes-Campos C, Elizondo RJ, Shoji KF, Contreras-Duarte S, Figueroa V, Velarde V, Jiang JX, Nualart F, Saez JC, Garcia MA. 2012. Glucose increases intracellular free Ca(2+) in tanycytes via ATP released through connexin 43 hemichannels. Glia 60:53–68.

O'Shea RD, Gundlach AL. 1991. Preproneuropeptide Y messenger ribonucleic acid in the hypothalamic arcuate nucleus of the rat is increased by food deprivation or dehydration. J. Neuroendocrinol. 3:11–14.

Owino S, Sanchez-Bretano A, Tchio C, Cecon E, Karamitri A, Dam J, Jockers R, Piccione G, Noh HL, Kim T, Kim JK, Baba K, Tosini G. 2018. Nocturnal activation of melatonin receptor type 1 signaling modulates diurnal insulin sensitivity via regulation of PI3K activity. J. Pineal Res. 64:10.

Pan A, Schernhammer ES, Sun Q, Hu FB. 2011. Rotating night shift work and risk of Type 2 diabetes: two prospective cohort studies in women. PLoS Medicine 8:e1001141.

Park JY, Pillinger MH, Abramson SB. 2006. Prostaglandin E2 synthesis and secretion: the role of PGE2 synthases. Clin. Immun. 119:229–240.

Pattaranit R, van den Berg HA. 2008. Mathematical models of energy homeostasis. J R Soc Interface 5:1119–1135.

Pérez JH, Tolla E, Dunn I, Meddle SL, Stevenson TJ. 2019. A comparative perspective on extraretinal photoreception. Trends in Endocrinology and Metabolism 30:39–53.

Petri I, Diedrich V, Wilson D, Fernandez-Calleja J, Herwig A, Steinlechner S, Barrett P. 2016. Orchestration of gene expression across the seasons: hypothalamic gene expression in natural photoperiod throughout the year in the Siberian hamster. Sci Rep. 6:29689.

Petri I, Dumbell R, Scherbarth F, Steinlechner S, Barrett P. 2014. Effect of exercise on photoperiod-regulated hypothalamic gene expression and peripheral hormones in the seasonal dwarf hamster *Phodopus sungorus*. PLoS One 9(3):e90253.Pfaff DW. 1968. Autoradiographic localization of radioactivity in rat brain after injection of tritiated sex hormones. Science 161:1355–1356.

Pfaff DW. 1999. *Drive: neurobiological and molecular mechanisms of sexual motivation*. Cambridge, MA: Bradford Book.

Pfaff DW, Know LM, Loose MD, Flanagan-Cato LM. 2008. Reverse engineering the lordosis behavior circuit. Horm. Behav. 54:347–354.

Piet R, Dunckley H, Lee K, Herbison AE. 2016. Vasoactive intestinal peptide excites GnRH neurons in male and female mice. Endocrinol. 157:3621–3630.

Pitman DL, Ottenweller JE, Natelson BH. 1988. Plasma corticosterone levels during repeated presentation of two intensities of restraint stress: chronic stress and habituation. Physiol. Behav. 43:47–55.

Platt JR. 1964. Strong inference: certain systematic methods of scientific thinking may produce much more rapid progress than others. Science 146:347–353.

Plomin R, Owen MJ, McGuffin P. 1994. The genetic basis of complex human behaviors. Science 264:1733–1739.

Poole JH. 1987. Rutting behavior in African elephants: the phenomenon of musth. Behaviour 102:283–316.

Posfai E, Schell JP, Janiszewski A, Rovic I, Murray A, Bradshaw B, Yamakaka T, Pardon T, El Bakkali M, Talon I, De Geest N, Kumar P, To SK, Petropoulos S, Jurisicova A, Pasque V, Lanner F, Rossant J. 2021. Evaluating totipotency using criteria of increasing stringency. Nature Cell Biol. 23:49–60.

Prendergast BJ, Cable EJ, Stevenson TJ, Onishi KG, Zucker I Kay LM. 2015. Circadian disruption alters the effects of lipopolysaccharide treatment on circadian and ultradian locomotor activity and body temperature rhythms of female Siberian hamsters. J. Biol. Rhythms 30:543–556.

Provencio I, Jiang G, De Grip WJ, Hayes WP, Rollag MD. 1998. Melanopsin: an opsin in melanophores, brain and eye. Proc. Natl. Acad. Sci. 95:340–345.

Pruessner JC, Dedovic K, Khalili-Mahani N, Engert V, Pruessner M, Buss C, Renwick R, Dagher A, Meaney MJ, Lupien S. 2008. Deactivation of the limbic system during acute psychosocial stress: evidence from positron emission tomography and functional magnetic resonance imaging studies. Biol. Psychiatry 63:234–240.

Radley J, Morilak D, Viau V, Campeau S. 2015. Chronic stress and brain plasticity: mechanisms underlying adaptive and maladaptive changes and implications for stress-related CNS disorders. Neurosci. Biobehav. Rev. 58:79–91.

Ranels HJ, Griffin JD. 2003. The effects of prostaglandin E2 on the firing rate activity of thermosensitive and temperature insensitive neurons in the ventromedial preoptic area of the rat hypothalamus. Brain Res. 964:42–50.

Ratajczak W, Niedzwiedzka-Rystwej P, Tokarz-Deptula B, Deptula W. 2018. Immunological memory cells. Cent. Eur. J. Immunol. 43:194–203.

Reddy AB, Cronin AS, Ford H, Ebling FJP. 1999. Seasonal regulation of food intake and body weight in the male Siberian hamster: studies of hypothalamic orexin (hypocretin), neuropeptide Y (NPY) and pro-opiomelanocortin (POMC). Eur. J. Neurosci. 11:3255–3264.

Reddy P, Zehring WA, Wheeler DA, Pirrotta V, Hadfield C, Hall JC, Rosbash M. 1984. Molecular analysis of the period locus in *Drosophila melanogaster* and identification of a transcript involved in biological rhythms. Cell 38:701–710.

Reed TE, Grotan V, Jenouvrier S, Sæther BE, Visser ME. 2013. Population growth in a wild bird is buffered against phenological mismatch. Science 340:488–491.

Refinetti R, Cornelissen G, Halberg F. 2007. Procedures for numerical analyses of circadian rhythms. Biol. Rhythm Res. 38:275–325.

Rehwinkel J, Gack MU. 2020. RIG-I-like receptors: Their regulation and roles in RNA sensing. Nat. Rev. Immunol. 20:537–551.

Renfree MB, Shaw G. 2000. Diapause. Ann. Rev. Physiol. 62:353–375.

Reppert SM, Henshaw D, Schwartz WJ, Weaver DR. 1987. The circadian-gated timing of birth in rats: disruption by maternal SCN lesions or by removal of the fetal brain. Brain Res. 403:398–402.

Reynolds WW, Casterlin ME. 1979. Behavioural fever in teleost fishes. Nature 259:41–42.

Richter CP. 1978. Dark-active rat transformed into light active rat by destruction of 24-hr clock: function of 24-hr clock and synchronizers. Proc. Natl. Acad. Sci. 75:6276–6280.

Riddle O, Bates RW, and Dykshorn SW. 1993. The preparation, identification and assay of prolactin—a hormone of the anterior pituitary. Am. J. Physiol. 105:191–216.

Roenneberg T, Aschoff J. 1990. Annual rhythm of human reproduction: I. biology, sociology or both? J. Biol. Rhythms 5:195–216.

Romero LM, Dickens MJ, Cyr N. 2009. The Reactive Scope Model—a new model integrating homeostasis, allostasis and stress. Horm. Behav. 55:375–389.

Rotherbauer M, Patel N, Gondola H, Siwetz M, Hupperz B, Ertl P. 2017. A comparative study of five physiological key parameters between four different human trophoblast-derived cell lines. Sci. Rep. 7:5892.

Rousseau K, Atcha Z, Cagampang FRA, Le Rouzic P, Stirland JA, Ivanov TR, Ebling FJP, Klingenspor M, Loudon ASI. 2002. Photoperiodic regulation of leptin resistance in the seasonally breeding Siberian hamster (*Phodopus sungorus*). Endocrinol. 143:3083–3095.

Ruddle NH, Akirav EM. 2009. Secondary lymphoid organs: responding to genetic and environmental cues in ontogeny and the immune response. J. Immunol. 183:2205–2212.

Rudic RD, McNamara P, Curtis A-M, Boston RC, Panda S, Hogenesch JB, FitzGerald GA. 2004. BMAL1 and CLOCK, two essential components of the circadian clock, are involved in glucose homeostasis. PLoS Biol. 2(11):e377.

Sacher GA, Staffeldt EF. 1974. Relation of gestation time to brain weight for placental mammals: implications for the theory of vertebrate growth. Amer. Nat. 108:593–615.

Saenz de Miera C, Monecke S, Bartzen-Sprauer J, Laran-Chich MP, Pevet P, Hazlerigg DG, Simonneaux V. 2014. A circannual clock drives expression of genes central for seasonal reproduction. Curr. Biol. 24:1500–1506.

Sapolsky RM, Krey LC, McEwen BS. 1984a. Stress down-regulates corticosterone receptors in a site-specific manner in the brain. Endocrinol. 114:287–292.

Sapolsky RM, Krey LC, McEwen BS. 1984b. Glucocorticoid-sensitive hippocampal neurons are involved in terminating the adrenocortical stress response. Proc. Natl. Acad. Sci. 81:6174–6177.

Sapolsky RM. 2004. *Why zebras don't get ulcers: an updated guide to stress, stress related diseases, and coping*. 3rd Rev. Ed. New York: Macmillan Publishers.

Schwanzel-Fukuda M, Pfaff DW. 1989. Origin of luteinizing hormone-releasing hormone neurons. Nature 338:161–164.

Seminara SB, Messager S, Chatzidaki EE, Thresher RR, Acierno JS, Shagoury JK, Bo-Abbas Y, Kuohung W, Schwinof KM, Hendrick AG, Zahn D, Dixon BA, Kaiser UB, Slaugenhaupt SA, Gusella JF, O'Rahilly S, Carlton MBL, Crowley WF, Aparicio SAJR, College WH. 2003. The PGR54 gene as a regulator of puberty. N. Engl. J. Med. 349:1614–1627.

Semon R. 1904. *Daw Mneme als erhaltendes Prinzip im Wechsel des organischen Geschehens*. Leipzig: W. Engelmann.

Sette A, Crotty S. 2021. Adaptive immunity to SARS-CoV-2 and COVID19. Cell 184:861–880.

Seyle HS. 1946. The general adaptation syndrome and the diseases of adaptation. J. Clin. Endocrinol. 6:117–230.

Schally AV, Arimura A, Kastin AJ, Matsuo H, Baba Y, Redding TW, Nair RMG, Debeljuk L, White WF. 1971. Gonadotropin-releasing hormone: one polypeptide regulates secretion of luteinizing and follicle-stimulating hormones. Science 173:1036–1038.

Schulkin J. 2004. *Allostasis, homeostasis, and the cost of physiological adaptation*. Cambridge, UK: Cambridge University Press.

Schwiening CJ. 2012. A brief historical perspective: Hodgkin and Huxley. J. Physiol. 590:2571–2575.

Seibert K, Zhang Y, Leahy K, Hauser S, Masferrer J, Perkins W, Lee L, Isakson P. 1994. Pharmacological and biochemical demonstration of the role of cyclooxygenase 2 in inflammation and pain. Proc. Natl. Acad. Sci. 91:12013–12017.

Shahbazi MN, Zernicka-Goetz M.2018. Deconstructing and reconstructing the mouse and human early embryo. Nature Cell. Biol. 20:878–887.

Sharp PJ, 1980. Female reproduction. In Epple A, Stetson MH (Eds.) *Avian endocrinology* (pp. 435–454). New York: Academic Press.

Sheeba V, Sharma VK, Chandrashekaran MK, Joshi A. 1999. Persistence of eclosion rhythm in *Drosophila melanogaster* after 600 generations in an aperiodic environment. Naturwissenschaften 86:448–449.

Siffre M. 1964. *Beyond time*. New York: McGraw-Hill.

Smith JM. 1978. *The evolution of sex*. Cambridge, UK: Cambridge University Press.

Smith MJ, Jiennes L, Wise PM. 2000. Localization of the VIP2 receptor protein on GnRH neurons in the female rat. Endocrinol. 141:4317–4320.

Smith LB, Walker WH. 2015. Hormone signaling in the testis. In Tony P, Anthony Z (Eds.) *Knobil and Neill's physiology of Reproduction* (pp. 637–690) 4th Ed. Academic Press.

Smith S. 2004. Anterior pituitary hormones: development of a bioassay leading to the discovery of prolactin. Physiol. Report 287:E813–E814.

Smyllie NJ, Pilorz V, Boyd J, Meng QJ, Saer B, Chesham JE, Maywood ES, Krogager TP, Spiller DG, Boot-Handford R, White MRH, Hastings MH, Loudon ASI. 2016. Visualizing and quantifying intracellular behavior and abundance of the core circadian clock protein Period2. Curr. Biol. 26:1880–1886.

Sohn JW, Elmquist JK, Williams KW. 2013. Neuronal circuits that regulate feeding behavior and metabolism. Trends Neurosci. 36:504–512.

Solnica-Krezel L, Sepich DS. 2012. Gastrulation: making and shaping germ layers. Ann. Rev. Cell Develop. Biol. 28:687–717.

Steel E, Hinde RA. 1964. Effect of exogenous estrogen on brood patch development of intact and ovariectomized canaries. Nature 202:718–719.

Stephan FK, Zucker I. 1972. Circadian rhythms in drinking behavior and locomotor activity of rats are eliminated by hypothalamic lesions. Proc. Natl. Acad. Sci. 69:1583–1586.

Sterling P, Eyer J. 1988. Allostasis: a new paradigm to explain arousal pathology. In Fisher S, Reason J (Eds.) *Handbook of life stress, cognition and health* (pp. 629–649). New York: Wiley.

Stevenson TJ, Alward BA, Ebling FJP, Fernald RD, Kelly A, Ophir A. 2018. Revisiting Krogh's principle and the value of the comparative approach. Policy Insights from the Behavioral and Brain Sciences 5:118–125.

Stevenson TJ, Ball GF. 2009. Anatomical localization of the effects of reproductive state, castration and social milieu on cells immunoreactive for gonadotropin-releasing hormone-I in male European starlings (*Sturnus vulgaris*). J. Comp. Neurol. 517:146–155.

Stevenson TJ, Ball GF. 2011. Information theory and the neuropeptidergic regulation of seasonal reproduction in mammals and birds. Proc. R. Soc. B 278:2477–2485.

Stevenson TJ, MacDougall-Shackleton SA, Hahn TP, Ball GF. 2012. Gonadotropin-releasing hormone plasticity: a comparative perspective. Front. Neuroendocrinol. 33:287–300.

Stevenson TJ, Prendergast BJ. 2013. Reversible DNA methylation regulates seasonal photoperiodic time measurement. Proc. Natl. Acad. Sci.110:16651–16656.

Stevenson TJ, et al., 2015. Disrupted seasonal biology impacts health, food security, and ecosystems. Proc. Royal Soc. B 282:1–10.Stewart JR, 1997. Morphology and evolution of the egg of oviparous amniotes. In Sumida SS, Martin KLM (Eds.) *Amniote origins* (pp. 291–326). San Diego, CA: Academic Press.

Sussman D, Pang EW, Jetly R, Dunkley BT, Taylor MJ. 2016. Neuroanatomical features in soldiers with post-traumatic stress disorder. BMC Neurosci. 17:13.

Taliesnik S, Sawyer CH. 1986. Activation of the CNS noradrenergic system may inhibit as well as facilitate luteinizing hormone release. Neuroendocrinol. 44:265–268.

Tan CL, Knight ZA. 2018. Regulation of body temperature by the nervous system. Neuron 98:31–48.

Tay MZ, Poh CM, Renia L, MacAry PA, Ng LFP. 2020. The trinity of COVID19: immunity, infection and intervention. Nat. Rev. Immunol. 20:363–374.

Tendler A, Bar A, Mendelsohn-Cohen N, Karin O, Kohanim YK, Maimon L, Milo T, Raz M, Mayo A, Tanay A, Alon U. 2021. Hormone seasonality in medical records suggests circannual endocrine circuits. Proc. Natl. Acad. Sci. 118:e2003926188.

Thackeray SJ, Sparks TH, Frederiksen M, Burthe S, Bacon PJ, Bell JR, Botham MS, Brereton TM, Bright PW. Carvalho L, Clutton-Brock T, Dawson A, Edwards M, Elliott JM, Harrington R, Johns D, Jones ID, Jones JT, Leech DI, Roy DB, Scott WA, Smith M, Smithers RJ, Winfield IJ, Wanless S. 2010. Trophic levels asynchrony in rates of phenological change for marine, freshwater and terrestrial environments. Global Change Biol. 16:3304–3313.

Thompson JM, Irgens LM, Skjaerven R, Rasmussen S. 2007. Placenta weight percentile curves for singleton deliveries. BJOG 114:715–20.

Tickell WLN. 1968. The biology of the great albatrosses, Diomedea exulahs and Diomedea epomophora. Antarctic Research Series 12:1–56.

Tolstoy I. 1981. *James Clerk Maxwell: a biography*. Chicago: University of Chicago Press.

Torrellas A, Guaza C, Borrell J, Borrell S. 1980. Adrenal hormones and brain catecholamine responses to morning and afternoon immobilization stress in rats. Physio. Behav. 26:129–133.

Tsatsoulis A, Fountoulakis S. 2006. The protective role of exercise on stress system dysregulation and comorbidities. Ann. NY Acad. Sci. 1083:196–213.

Tschop M, Smiley DL, Heiman ML. 2000. Ghrelin induces adiposity in rodents. Nature 407:908–913.

Turek FW, Joshu C, Kohsaka A, Lin E, Ivanova G, McDearmon E, Laposky A, Losee-Olson S, Easton A, Jensen DR, Eckel RH, Takahashi JS, Bass J. 2005. Obesity and Metabolic syndrome in circadian clock mutant mice. Science 308:1043–1045.

Underwood H. 1989. The pineal and melatonin: regulators of circadian function in lower vertebrates. Experientia 45:914–922.

Vaisse C, Clement K, Guy-Grand B, Froguel P.1998. A frameshift mutation in human MC4R is associated with a dominant form of obesity. Nat. Genet. 20:113–114.

Van der Beek EM, Horvath TL, Wiegant VM, Van den Hurk R, Buijs RM. 1997. Evidence for a direct neuronal pathway from the suprachiasmatic nucleus to the gonadotropin-releasing hormone system: combined tracing and light and electron microscopic immunocytochemical studies. J. Comp. Neurol. 384:569–579.

Valent P, Groner B, Schumacher U, Superti-Furga G, Busslinger M, Kralovics R, Zielinski C, Penninger JM, Kerjaschki D, Stingl G, Smolen JS, Valenta R, Lassman H, Kovar H, Jager U. Kornek G, Muller M, Sorgel F. 2016. Paul Ehrlich (1854–1915) and his contributions to the foundation and birth of translational medicine. J. Innate. Immun. 8:111–120.

Vane JR, Botting RM. 2003. The mechanism of action of aspirin. Thromb. Res. 110:255–258.

Vaughn LK, Bernheim HA, Kluger MJ. 1974. Fever in the lizard *Dipsosaurus dorsalis*. Nature 252:473–474.

Vitaterna MH, King DP, Chang AM, Kornhauser JM, Lowrey PL, McDonald JD, Dove WF, Pinto LH, Turek FW, Takahashi JS. 1994. Mutagenesis and mapping of a mouse gene, Clock, essential for circadian behaviour. Science 264:719–725.

von Uexküll J. 2010 [1934]. *A foray into the worlds of animals and humans: with a theory of meaning*. Translated by Joseph D. O'Neil. Minneapolis/London: University of Minnesota Press.

Wade N. 1981. *The Nobel duel*. New York: DoubleDay Anchor Press.

Warner A, Jethwa PH, Wyse CA, I'Anson H, Brameld JM, Ebling FJP. 2010. Effects of photoperiod on daily locomotor activity, energy expenditure, and feeding behavior in a seasonal mammal. Am. J. Physiol. Reg. Integ. Comp. Physiol. 298:1409–1416.

Weaver ICG, Cervoni N, Champagne FA, D'Alessio AC, Sharma S, Seckl JR, Dymov S, Szyf M, Meaney MJ. 2004. Epigenetic programming by maternal behavior. Nat. Neurosci.7:847–854.

Wetendorf M, DeMayo FJ. 2012. The progesterone receptor regulates implantation, decidualization, and glandular development via a complex paracrine signaling network. Mol. Cell. Endocrinol. 357:108–118.

Wichmann S, Kirschbaum C, Bohme C, Petrowski K. 2017. Cortisol stress response in post-traumatic stress disorder, panic disorder, and major depressive disorder patients. Psychoneuroendocrinol. 83:135–141.

Williams WP, Kriegsfeld LJ. 2012. Circadian control of neuroendocrine circuits regulating female reproductive function. Front. Endocrinol. 3.

Williams WP III, Jarjisian SG, Mikkelsen JD, Kriegsfeld LJ. 2011. Circadian control of kisspeptin and a gated GnRH response mediate the preovulatory luteinizing hormone surge. Endocrinol. 152:595–606.

Wilson A, Trumpp A. 2006. Bone-marrow haematopoietic-stem-cell niches. Nat. Rev. Immunol. 6:93–106.

Wilson DM. 1961. The central nervous control of flight in a locust. J. Exp. Biol. 38:471–479.

Wooding P, Burton G. 2008. *Comparative placentation: structures, functions and evolution*. Cham, Switzerland: Springer.

Wulff K, Gatti S, Wettstein JG, Foster RG. 2010. Sleep and circadian rhythm disruption in psychiatric and neurodegenerative disease. Nat. Rev. Neurosci. 11:589–599.

Wyse CA, Morales CAC, Ward J, Lyall D, Smith DJ, Mackay D, Curtis AM, Bailey MES, Biello S, Gill JMR, Pell JP. 2018. Population-level seasonality in cardiovascular mortality, blood pressure, BMI and inflammatory cells in UK Biobank. Ann. Med. 50:410–419.

Xiao XQ, Grove KL, Lau SY, McWeeney S, Smith MS. 2005. Deoxyribonucleic acid microarray analyses of gene expression pattern in the arcuate nucleus/ventromedial nucleus of hypothalamus during lactation. Endocrinol. 146:4391–4398.

Yamanaka Y, Motoshima H, Uchida K. 2019. Hypothalamic-pituitary-adrenal axis differentially responses to morning and evening psychological stress in healthy subjects. Neuropsychopharmacol. Rep. 39:41–47.

Yang S, Liu A, Weidenhammer A, Cooksey RC, McClain D, Kim MK, Aguilera G, Abel ED, Chung JH. 2009. The role of mPer2 clock gene in glucocorticoid and feeding rhythms. Endocrinol. 150:2153–2160.

Yaswen L, Diegl N, Brennan MB, Hochgeschwender U. 1999. Obesity in the mouse model of proopiomelanocortin deficiency responds to peripheral melanocortin. Nat. Med. 5:1066–1070.

Yeo GS, Farooqi IS, Aminian S, Halsall DJ, Stanhope RG, O'Rahilly S. 1998. A frameshift mutation in MC4R associated with dominantly inherited human obesity. Nat. Genet. 20:111–112.

Yeo GS, Heisler LK. 2012. Unraveling the brain regulation of appetite: lessons from genetics. Nat. Neurosci. 15:1343–1349.Yoshimura T. 2013. Thyroid hormone and seasonal regulation of reproduction. Front. Neuroendocrinol. 34:157–166.

Yoo S, Cha D, Kim S, Jiang L, Cooke P, Adebesin M, Wolfe A, Riddle R, Aja S, Blackshaw S. 2020. Tanycyte ablation in the arcuate nucleus and median eminence increases obesity susceptibility by increasing body fat content in male mice. Glia 68:1987–2000.

Yoo SH, Yamazaki S, Lowrey PL, Shimomura K, Ko CH, Buhr ED, Siepka SM, Hong HK, Oh WJ, Yoo OJ, Menaker M, Takahashi JS. 2004. PERIOD2:LUCIFERASE real-time reporting of circadian dynamics reveals persistent circadian oscillations in mouse peripheral tissues. Proc. Natl. Acad. Sci. 101:5339–5346.Zernicka-Goetz M, Highfield R. 2020. *The dance of life*. London: Penguin Random House.

Zhang Y, Proenca R, Maffei M, Barone M, Leopold L, Friedman J. 1994. Positional cloning of the mouse obese gene and its human homologue. Nature 372:425–432.

Zhao S, Kriegsfeld LJ. 2009. Daily changes in GT1-7 cell sensitivity to GnRH secretagogues that trigger ovulation. Neuroendocrinol. 89:448–457.

Zhao ZD, Yang WZ, Gao C, Fu X, Zhang W, Zhou Q, Chen W, Ni X, Lin JK, Yang J, Xu XH, Shen WL. 2017. A hypothalamic circuit that controls body temperature. Proc. Natl. Acad. Sci. 114:2042–2047.

Zigman JM, Jones JE, Lee CE, Saper CB, Elmquist JK. 2005. Expression of ghrelin receptor mRNA in the rat and the mouse brain. J. Comp. Neurol. 494:528–548.

Zigman JM, Nakano Y, Coppari R, Balthasar N, Marcus JN, Lee CE, Jones JE, Deysher AE, Waxman AR, White RD, Williams TD, Lachey JL, Seeley RJ, Lowell BB, Elmquist JK. 2005. Mice lacking ghrelin receptors resist the development of diet-induced obesity. J. Clin. Invest. 115:3564–3572.

Zimmerman NH, Menaker M. 1979. The pineal gland: a pacemaker within the circadian system of the house sparrow. Proc. Natl Acad. Sci. 76:999–1003.

Zucker I, Boshes M. 1982. Circannual body weight rhythms of ground squirrels: role of gonadal hormones. Am. J. Physiol. 243:R546–R551.

Index

For the benefit of digital users, indexed terms that span two pages (e.g., 52–53) may, on occasion, appear on only one of those pages.

Tables and figures are indicated by *t* and *f* following the page number